Sharing and Hiding Religious Knowledge in Early Judaism,
Christianity, and Islam

Judaism, Christianity, and Islam – Tension, Transmission, Transformation

Edited by
Patrice Brodeur, Alexandra Cuffel,
Assaad Elias Kattan, and Georges Tamer

Volume 10

Sharing and Hiding Religious Knowledge in Early Judaism, Christianity, and Islam

Edited by
Mladen Popović, Lautaro Roig Lanzillotta,
and Clare Wilde

DE GRUYTER

ISBN 978-3-11-064373-2
e-ISBN (PDF) 978-3-11-059660-1
e-ISBN (EPUB) 978-3-11-059366-2
ISSN 2196-405X

Library of Congress Control Number: 2018943762

Bibliographic information published by the Deutsche Nationalbibliothek
The Deutsche Nationalbibliothek lists this publication in the Deutsche Nationalbibliografie;
detailed bibliographic data are available in the Internet at http://dnb.dnb.de.

© 2020 Walter de Gruyter GmbH, Berlin/Boston
This volume is text- and page-identical with the hardback published in 2018.
Typesetting: Integra Software Services Pvt. Ltd.
Printing and binding: CPI books GmbH, Leck

www.degruyter.com

Acknowledgments

This collection of essays arises from the international conference on 22–24 April 2015 organized by the Department of Jewish, Christian, and Islamic Origins at the Faculty of Theology and Religious Studies of the University of Groningen: "Sharing and Hiding Religious Knowledge: Strategies of Acculturation and Cultural Resistance in Early Jewish, Christian, and Islamic Traditions." The organizers are very grateful to the Royal Netherlands Academy of Arts and Sciences and the Nicolaas Mulerius Fund (University of Groningen) for awarding the necessary grants that made the conference possible. The resulting essays have been thoroughly reworked for inclusion into this volume. We wish to thank the series editors of De Gruyter Publishers for accepting this volume for publication into their series Judaism, Christianity, and Islam – Tension, Transmission, Transformation.

Contents

Mladen Popović, Lautaro Roig Lanzillotta, and Clare Wilde
Introduction —— 1

Eleanor Robson
1 Do Not Disperse the Collection! Motivations and Strategies for Protecting Cuneiform Scholarship in the First Millennium BCE —— 8

Mladen Popović
2 Multilingualism, Multiscripturalism, and Knowledge Transfer in the Dead Sea Scrolls and Graeco-Roman Judaea —— 46

Jacques van Ruiten
3 Sharing and Hiding Religious Knowledge in the Book of Jubilees —— 72

Katell Berthelot
4 The Torah Between Revelation and Concealment in Rabbinic Traditions Pertaining to the Conquest of the Land of Canaan —— 85

Delfim F. Leão
5 Alexandria, Diaspora, *Politeuma* and *Patrioi Nomoi*: The Sharing and Hiding of Jewish Identity —— 106

Lautaro Roig Lanzillotta
6 Ancient Greek Patterns of Knowledge Transmission and their Continuity in Gnostic Esotericism —— 121

George van Kooten
7 The Sign of Socrates, the Sign of Apollo, and the Signs of Christ: Hiding and Sharing Religious Knowledge in the Gospel of John – A Contrapuntal Reading of John's Gospel and Plato's Dialogues —— 145

Clare Wilde
8 "They Wish to Extinguish the Light of God with Their Mouths" (Qurʾān 9:32): A Qurʾānic Critique of Late Antique Scholasticism? —— 171

Paul E. Walker
9 Techniques for Guarding and Restricting Esoteric Knowledge in the Ismaili *Da'wa* during the Fatimid Period —— 186

Author Index —— 199

Sources Index —— 204

Mladen Popović, Lautaro Roig Lanzillotta, and Clare Wilde
Introduction
Sharing and Hiding Religious Knowledge in Early Jewish, Christian, and Islamic Traditions

Knowledge in Antiquity was cherished as a scarce good and its character and transmission tainted with an esoteric allure. The production and cultivation of knowledge not only took place within the limited circle of the "initiated"; its diffusion was also channelled through the close relationship teacher-disciple.

The esoteric aspect plays a central role in scholarly, scribal, religious and philosophical contexts. Knowledge was not only intended for a limited group of followers; it also seemed to provide a higher form of consciousness that not everyone was willing or able to bear. If from an existential perspective, knowledge provides individuals with a holistic framework to supersede a fragmented reality, from a social viewpoint, it provides them with the means to advance in the social hierarchy. On the one hand, possessing or lacking knowledge determines social status; on the other, sharing or hiding knowledge is used in strategies of inclusion and exclusion that are highly productive both at the micro (within religious communities themselves) and the macro levels (within multicultural societies at large).

Whether religious knowledge could or should be shared with others or, instead, kept to oneself was one of the central issues by which Jews, Christians, and Muslims defined themselves in relation to each other and the world around them. The formative stages of each of these traditions were characterised by a wide diversity of attitudes towards, and means of, knowledge sharing and hiding. Although this sharing and hiding could be textual, such as the Wisdom literature in the Hebrew Bible or the revelatory knowledge throughout Jewish tradition, this volume focuses on the cultural encounter between Jews, Babylonians, Greeks, Romans, and others. With whom was religious knowledge shared or from whom was it hidden?

Although the hiding of knowledge connotes an active strategy of concealment, it does not preclude the possibility of simply not sharing certain things. And the refusal to share aspects of religious knowledge may also illuminate the nature of specific cultural encounters. Is the transmission of knowledge geared toward internal consumption or is it shared with outsiders? With respect to Judaism, these questions have emerged in recent discussions about the position of rabbinic cultures within the Roman Empire: were they part of it or resistant to it? The sharing and hiding of knowledge also relates to processes of acculturation and cultural resistance. How, for example, should the cultural transfer of knowledge and multilingualism in bodies of learning that are seemingly culturally resistant, such as the Dead Sea Scrolls, be understood? How does this relate, from a comparative perspective, to the Epicurean library from Herculaneum, which belonged to Philodemus, a philosopher who came from nearby Gadara, but seemingly from an entirely different culture?

The tension between esoteric and open, namely between hiding and sharing knowledge, is at the core of the early Christian movement. While proto-orthodox Christians claimed an open and public notion of religious knowledge that was accessible to everyone, heterodox groups conceived it in esoteric terms, limiting it to a small group of initiates. Both Christian views of knowledge reflect divergent attitudes towards the cultural and religious domination of the Graeco-Roman society. On the one hand, the universalism of proto-orthodoxy overtly rejects the esoteric character of some late antique religious experience and might therefore be described as "cultural resistance"; on the other hand, heterodox perceptions of religious knowledge clearly show a continuity with the surrounding world. They thus provide a good example of acculturation, in which Christians accommodated to the cultural standards of the Graeco-Roman religious experience.

The quest for knowledge and learning is, according to a canonical *ḥadīth*, enjoined on every Muslim. The importance of knowledge and learning is also in the Qur'ān which states explicitly: "Say: 'Are they equal – those who know and those who know not?' Only those with understanding will take the admonishment to heart" (Qur'ān 39:9). The quest for and spread of knowledge and learning became one of the important characteristics of the emerging religion and civilization.

Two kinds of knowledge developed, both of which were based, to a certain degree, on the Qur'ān and Sunna. One focused on the apparent, manifest, exoteric sense; the other, its hidden, inner, esoteric sense. The former is exemplified by the caliph 'Uthmān sending copies of the vulgate text of the Qur'ān to the metropoles of the new empire. Even if this account is not historically reliable, it is at least symbolically true: it highlights the text-centric nature of religious learning in the Arab-Islamic civilisations. Whatever geographical centres of learning may have existed, the Qur'ān became the ideological centre of Islam's religious learning, shaping many other kinds of learning as well, including the Arabic language itself. The mystical tradition, on the other hand, emphasised knowledge of the inner self, or internal knowledge, for which a long period of spiritual and physical exercises under the strict guidance of an adept teacher was nearly always needed.

This collection of essays examines the processes and reasons for sharing and hiding religious knowledge through cuneiform texts, as well as in Jewish, Christian, and Islamic contexts. It begins with Eleanor Robson's assessment of the social and technical aspects of protecting written productions in Assyrian and Babylonian cuneiform texts from the first millennium BC. Her study emphasizes the social dimensions of esotericism, since sharing or hiding knowledge become means of inclusion and exclusion in religious communities and society at large. She argues that this period witnessed a dramatic change in the function and character of written products: transitioning from simple aide memoires to precious objects in their own right. Colophons reflect this well, since they now include, besides the date and purpose of the writing, and the dispositions of the clay tablets, injunctions intended to protect both the tablets and their contents. Assyrian examples include warnings to protect the texts from "the uninitiated," reflecting the scribes' desire to protect their knowledge from the political centre and its attempts to create large tablet collections in order to monopolize access to learning. This scribal protection suggests that possessing or lacking knowledge determined social status; thus, the acquisition of knowledge became a means to advance in the social hierarchy. Similar protections can be seen in sixth century Babylonian texts, but, Robson argues, in this case, scribes were threatened with loss of income and status within the urban community. Rather than protecting the secrecy or esoteric nature of the texts, the scribes' protection was more intended to vouchsafe their position, status and income.

The next chapters (three to six) turn to the world of Judaism. Mladen Popović focuses on multilingualism, "multiscripturalism," and knowledge transfer in the Dead Sea Scrolls. He conceives of the people behind the Dead Sea Scrolls as a textual community and rejects the sociological assumption that we are dealing with a small, isolated, marginal community at the site of Khirbet Qumran. On the basis of insights from recent sociolinguistic research on multilingualism and minority languages from a centre-periphery dynamics perspective, this study argues that the manuscripts from Qumran should no longer be framed in centre-periphery terms, with Qumran labeled as deviating from a standard norm. The multilingual evidence from Qumran and the use of several scripts is contextualized with other manuscript finds in the Judaean desert. Arguing against the thesis of Qumran as an isolated pocket of monolingual language ideology, he presents sharing and hiding as forms of social interaction. The emphasis is not on the contents of the secret, but on the significance of the concealment of knowledge. In Graeco-Roman Judaea, Popović argues, specific strategies for sharing or hiding learned knowledge were in operation by means of multilingualism, multiscripturalism, and scholarly literacy.

In the following chapter, Jacques van Ruiten analyses the Book of Jubilees and assesses its approach to the notions of sharing and hiding religious knowledge. The Book of Jubilees is a good example of the use of secrets in order to attract and repel potential participants by limiting those who are entitled to receive and transmit it. Knowledge that can be shared includes knowledge about halakhic and calendrical affairs, and apotropaic knowledge against the influence of evil spirits. Not everyone, however, has access to this knowledge: even if it is not necessarily secret, this sort of knowledge originates in heaven and is meant for the chosen sons only. It proceeds from the angels and is transmitted to Enoch, via Noah, Abraham, Jacob, Levi and Moses. Knowledge, however, that should be hidden includes the astrological knowledge of the Chaldeans, which is connected to the teachings of the fallen angels, the watchers. Given that this knowledge comes from outside of Israel and consequently represents a potential threat, it should be kept concealed. Here, a secret (shared) knowledge serves to cement mutual relations within the group: by participating in the divine origin of knowledge and transmitting it within the closed group. On the other hand, the rejection of alien knowledge is based on the commandment to separate from the nations and to avoid social contacts. This rejection of foreign knowledge is expressed in the idea that Moses got his education not at the Egyptian court but from his father Amram. This protective attitude is rooted in the election of Israel by God, which implies the exclusion of everything that could threaten this exclusive relationship.

From Jubilees we next turn to the position of rabbinic cultures in the Roman empire. Drawing on examples from rabbinic traditions pertaining to the land of Canaan, Katell Berthelot explores how the rhetoric of sharing and hiding religious knowledge may disguise processes of acculturation and cultural resistance. She analyses the ways in which rabbinic texts speak about both the revelation and the concealment of the Torah, paying attention to the rhetoric of each text in order to highlight the exegetical, theological and ethical issues that determine whether the knowledge of the Torah should or should not be shared with Gentiles. She demonstrates how, even if the argumentation varies from text to text, it is possible to recover a general logic behind them. They are all based on the Hebrew Bible and Moses's commandment to Joshua and Israel slightly before the conquest of Canaan. And, situated as they are in the historical context of the Roman Empire, these texts can be seen as rabbinic attempts to resist imperial domination.

In the following chapter, Delfim Leão explores the ways in which distinctive identities were safeguarded in highly cosmopolitan Alexandria. Against the ethnic, cultural and linguistic fusion on which Alexandria's success was grounded, he distinguishes two types of self-preservation. On the one hand, Macedonian and Greek communities who were close to the governing elite managed to ground their behavior in the so-called concept of *politikoi nomoi*, rules which derived from a common political and cultural identity. The Jewish community, however, also managed to obtain the right to "live according to their ancestral laws." Arguing that the Greek translation of the Torah (the Septuagint) helped in this differentiation, he explains that this sacred text had a similar status to that of the Greek *nomoi*, thus allowing the development of a legal *koine* on which to base daily life and to deal with private conflicts. The author compares the legal situation of the Jews and that of the Greeks, taking as reference the Jewish *politeuma* of Alexandria, whose existence, if historically accepted, exemplifies how Jews from the Diaspora could organize themselves into stable communities, from a religious, political and legal standpoint.

The following chapters (seven and eight) turn to the world of early Christianity, focusing on both the cultural context and the continuity and discontinuity of knowledge production and transmission among early Christians. According to Lautaro Roig Lanzillotta, Gnosticism perpetuated ancient Greek patterns of knowledge transmission well rooted in the ancient Mediterranean world. When protecting their knowledge and reserving it for a small group of initiates, Gnostics were in fact continuing a long and well-established tradition in Graeco-Roman philosophical schools, known since the Pre-Socratics. With the promise of giving access to deeper truths hidden under the surface of ordinary things, Gnostic secret knowledge intended to provide a holistic approach to reality, a charter or framework that helped people to overcome the perception of a fragmented reality. The higher level of consciousness attained by this special knowledge allowed individuals to transform both their person and their life, thus providing consistence and coherence to their worldview. The tension between secret and revelation was intended to protect this knowledge from outsiders, while also endowing those possessing it with a higher esteem. Against current views that conceive of Gnostic esotericism as a sectarian development of first-century Judaism, the author shows that Gnostics were following the pattern provided by all philosophical schools of the period, associations, and clubs or *collegia* in their cultural context, namely the Graeco-Roman world. In so doing, Gnostic groups were not "accommodating" their views to the Graeco-Roman setting. As part and parcel of the ancient world they did not need to adapt to a different culture, but simply expressed their views and beliefs according to the cultural standards of the world they lived in.

In the next chapter, George van Kooten focuses on the tension between concealing and revealing and the ability to create meaning for those capable of decoding it, explaining it through the Gospel of John's strategies regarding the notions of sharing and hiding religious knowledge. He argues that the tension between "neither telling nor concealing" is shaped by Heraclitus (B 93 DK, "the Lord whose prophetic shrine is at Delphi neither tells nor conceals, but signifies"), and runs through the Johannine narrative, together with Heraclitean terminology. On the one hand, despite being the Logos, Jesus does not speak clearly (as reflected in the response of the audience that Jesus's word is a λόγος σκληρός, in 6.60, a hard word, difficult to understand); on the other hand, Jesus did not completely conceal (himself) either. Based on Plutarch's reflections on Heraclitus's fragment in *De Pythiae oraculis* (404D–E), the author states that John may be understood as a narrative developing the Heraclitean axiom (in line with the Chaeronean) of a God "neither telling nor concealing, but signifying." The transition from the indirect, diffuse circumlocution of λόγοι σκληροί, παροιμίαι and σημεῖα to plain speech at the crucial moment of the last Symposium is a clear instance of sharing and hiding in one of the constitutive, canonical writings of Christianity, the Gospel of John.

The final chapters (9 and 10) introduce us to the world of Islam and the way knowledge and its transmission was conceived of in this environment. Clare Wilde explores a possible qur'ānic critique of Late Antique scholasticism in the verse "They wish to extinguish the light of God with their mouths" (Qur'ān 9:32). Corruption of scripture has frequently been understood as Jewish (or Christian) scribal manipulation of a written text only. The Qur'ān, however, also alludes to oral distortions likely to appear in the transmission of its text: it is with their mouths that Jews say that Uzayr is the son of God, and Christians say that the Messiah is (Qur'ān 9:30). This chapter provides an analysis of the way in which the Qur'ān refers to both written and oral scriptural corruptions in the light of trends in Late Antique circles. It claims that when attacking the desire to "extinguish the light of Gods with their mouths," the Qur'ān parallels certain criticism of (Hellenic) scholasticism found in Syriac Christian monastic literature. Following Adam Becker's study of the School of Nisibis (on the current Turkish/Syrian border) and the parallels between the East Syrian schools and the Babylonian Rabbinic academies, the author examines the possibility of qur'ānic knowledge of the debates over scholasticism in rabbinic and monastic circles.

Finally, Paul Walker discusses the strategies for keeping esoteric knowledge in the Ismaili Da'wa during the Fatimid period and before, as well as the different ways of approaching the distinction between literal and allegorical meaning of scripture in Islam. He explains that there are numerous techniques to restrict the access to knowledge: for example, the oath administered to each adept who wishes to join the cause; the strict prohibition against novices' sharing knowledge received without permission; or the obligation of paying tithes and fees, which function as a test of the new member's sincerity, as only after such payments is the agent of the da'wa allowed to impart esoteric knowledge. As necessary background for understanding Ismaili esotericism, this chapter also explores Islamic discussions of what knowledge can and cannot be revealed openly. Taking its starting point from Qur'ān 3:7, which admits that the sacred book contains verses that are ambiguous, it delves into the distinction between literal meaning (the exoteric dimension, ẓāhir) and the truth behind the outward sense (the esoteric, bāṭin). While common understanding claims that only God can know the meaning of ambiguous verses requiring interpretation (ta'wīl), philosophers thought that elite scholars had the necessary knowledge for such interpretations. They must not, however, share this with the general public. According to the Ismaili Shi'a, knowledge of the inner, esoteric meaning of scripture, as with the ambiguous verses of the Qur'ān, falls to the imams only, is concealed and thus not available to ordinary Muslims. Averroes does not agree, however, and claims that allegorical interpretation of scripture can be reached rationally. For the Ismailis, as explained by their principal authorities, the matter is more complicated. There are layers of explanation. All items of doctrine in scripture and the Law can have symbolic value. There may be no logical connection between the symbol and what is symbolized by it. Here is knowledge that is truly esoteric.

Eleanor Robson
1 Do Not Disperse the Collection! Motivations and Strategies for Protecting Cuneiform Scholarship in the First Millennium BCE

1.1 Introduction

By the early first millennium BCE, cuneiform culture was fighting a long, slow battle against obsolescence. Alphabetic scripts from the Levant, comprising just a few dozen characters, were easy to memorise and straightforward to use. By contrast the venerable family of cuneiform scripts had acquired multiple layers of complexity over more than two millennia of use in Babylonia, Assyria, and their spheres of influence. A functionary of the Assyrian Empire in the eighth or seventh centuries BCE minimally needed to master nearly 100 cuneiform signs, with around 35 logographic and over 80 syllabic values, in order to read everyday imperial correspondence.[1] This was a significant intellectual burden, which even the governor of the Babylonian city of Ur sought to be relieved of, asking Sargon II in c.800 BCE: "if it is acceptable to the king, let me write and send my messages to the king in Aramaic."[2] The king refused, citing not practical reasons but protocol and his own personal preference: it was "an established regulation" that royal correspondence must be in Akkadian cuneiform.[3]

Anyone with pretensions to learning required perhaps five times or more than that range of reading knowledge, not only in the vernacular Semitic language Akkadian but also in the literary isolate Sumerian,[4] acquired through years

---

[1] Greta Van Buylaere, "A Palaeographic Analysis of Neo-Assyrian" (PhD diss., University of Udine, 2009).
[2] ⸢ki-i⸣ [IGI] ⸢LUGAL⸣ mah-ru ina ŠÀ si-ip-ri | [^kur)]ár-⸢ma⸣-[a-a lu]-⸢us⸣-pi-ir-ma (Manfred Dietrich, *The Neo-Babylonian Correspondence of Sargon and Sennacherib*, State Archives of Assyria 17 [Helsinki: Helsinki University Press, 2003], no. 2 obv. 15–16).
[3] mi-nam-ma ina ši-pir-ti | ak-ka-da-at-tu la ta-šaṭ-ṭar-ma | la tu-šeb-bi-la kit-ta ši-pir-tu | šá ina ŠÀ-bi ta-šaṭ-ṭa-ru | ki-i pi-i a-gan-ni-tim-ma i-da-at | lu-ú šak-na-at "Why do you not write and send Akkadian in messages? Truly, the message that you write in it must be according to these conventions. It really is an established regulation." (Dietrich, *The Neo-Babylonian Correspondence*, no. 2 obv. 16–20).
[4] Eleanor Robson and Greta Van Buylaere, "Assyrian-Babylonian Scholarly Literacies" (unpublished manuscript).

of painstaking copying and rote memorisation, studying under a master scholar.[5] Even the simplest cuneiform texts were a challenge to read but most people with a reasonable degree of functional literacy would probably also have been able to muddle through a royal inscription or a passage from a narrative literary text such as The Epic of Gilgamesh, as these genres mostly used simple spelling conventions. However, mastering genres such as divination, healing, incantation, and ritual required further specialised learning: not only technical vocabulary but also highly context-specific spellings.[6]

Take for example the simple word *šumma*, "if." An Assyrian imperial bureaucrat could choose to write this as *šum-ma*, *šúm-ma* or possibly *šum₄-ma* (where acute and grave accents and subscript numerals are the modern convention for disambiguating homophonous cuneiform signs in alphabetic transliteration). He would have been expected to recognise all three alternatives when reading.[7] However, a scholar of terrestrial or celestial omens, a healer looking up medical recipes, or a performer of incantations and rituals also had to be conversant with the logographic writings BE and U₄ – which represent the whole word in one short sign – as well as the Sumerian *tukum-bi*, written with a long sequence comprising the signs ŠU, GAR, TUR, LAL, and BI.[8] Conversely, in everyday contexts the noun *amēlu*, "man", was almost invariably written with the simple logogram LÚ. But scholarly genres could in addition substitute it with NA, syllabic spellings such as *a-me-lu*, *a-mé-lu*, *a-me₈-lu₄* or *à-me₈-lú*, or even the elaborate logogram LÚ.U₁₈.LU.

In the light of these highly differentiated cuneiform literacies then, what are we to make of the fact that some copyists of scholarly works were apparently obsessed with the thought that others might steal their knowledge? From at least the late second millennium BCE, and regularly from the eighth century BCE onwards, we find injunctions to secrecy, and against loss and theft, on a wide variety of tablets written by a range of different people.[9] For instance, in 701 BCE

---

5 Petra D. Gesche, *Schulunterricht in Babylonien im ersten Jahrtausend v. Chr.*, Alter Orient und Altes Testament 275 (Münster: Ugarit, 2000); Eleanor Robson, "The Production and Dissemination of Scholarly Knowledge," in *The Oxford Handbook of Cuneiform Culture*, ed. Karen Radner and Eleanor Robson (Oxford: Oxford University Press, 2011), 557–76, at 562–69.
6 Niek Veldhuis, "Levels of Literacy," in *The Oxford Handbook of Cuneiform Culture*, ed. Karen Radner and Eleanor Robson (Oxford: Oxford University Press, 2011), 68–89.
7 Data from the Neo-Assyrian glossary of the State Archives of Assyria online http://oracc.org/saao/akk-x-neoass, accessed 8 August 2016.
8 Data from the Standard Babylonian and Sumerian glossaries of the Corpus of Ancient Mesopotamian Scholarship http://oracc.org/cams/gkab/akk-x-stdbab and http://oracc.org/cams/gkab/sux, accessed 8 August 2016.
9 For Middle Babylonian and Middle Assyrian examples see, e.g., Hermann Hunger, *Babylonische und Assyrische Kolophone* (Kevelaer: Butzon & Bercker; Neukirchen-Vluyn: Neukirchener,

in the Assyrian provincial town of Huzirina, apprentice scribe Nabu-rehtu-uṣur copied out the literary comedy now known as The Poor Man of Nippur, enjoining:

> Whoever takes away (this tablet), may the god Ea take him away! At the command of the god Nabu, who lives in the Ezida temple, may he have no descendants, no offspring!
>
> Do not take away the tablets! Do not disperse the collection! Taboo of the god Ea, king of the Abyss.[10]

Over half a millennium later, in the southern Babylonian city of Uruk, the young Anu-aba-uter calculated a table of expected lunar eclipses for his father Anu-belšunu, a *kalû*-lamenter. Dating his tablet to the ancient equivalent of April 191 BCE, he admonished:

> Whoever fears the gods Anu, Ellil and Ea [shall not take] it away by theft(?). Ephemeris, wisdom of Anu-ship, secret of the [great] gods, treasure of the scholars. The learned may show [the learned]; the unlearned may not [see. Taboo] of Anu, Ellil [and Ea, the great gods].[11]

Who were these putative thieves, the "unlearned" yet highly cuneiform-literate rogues who would risk the wrath of the gods in order to gain access to such texts? Given the huge amount of time and intellectual labour that the scholars themselves had personally invested in the acquisition of sufficient expertise to comprehend learned writings, they cannot possibly have imagined that a casual reader could make any sense of such a tablet if they had found one dropped in the street. Yet the threat was real enough for genuine concern to be expressed again and again over millennia. This paper attempts to answers the conundrum of the perceived vulnerability of this intrinsically impenetrable knowledge system.[12]

1968), nos. 40, 50; Alan Lenzi, *Secrecy and the Gods: Secret Knowledge in Ancient Mesopotamia and Biblical Israel* (Helsinki: The Neo-Assyrian Text Corpus Project, 2008), 216–19.

10 ša IR ᵈ60 lit-bal-šú | ina qí-bit ᵈMUATI a-šib É.ZI.DA | a-a GÁL⁻ˢⁱ NUNUZ-šú na-an-nab-šú ṭup-pi la ⌈ta-ta⌉-bil | ⁱᵐGÚ.[LÁ] la ta-par-ra-ru | [ik]-⌈kib⌉ ᵈ60 LUGAL ABZU (Oliver R. Gurney and Jacob J. Finkelstein, *The Sultantepe Tablets, Volume I* [London: British Institute of Archaeology at Ankara, 1957], no. 38 rev. ii 11–13, 16–18).

11 pa-lih 21 50 u 40 ina šur⁈⌈qa⌉⁈ [la TÙM]-šú | a-ru-ú né-me-qí ᵈ60-ú-tú ⌈AD.HAL DINGIR⌉.[MEŠ GAL.MEŠ] | MÍ.ÙRI ˡᵘ́um-man-nu ˡᵘ́ZU⁻ᵘ́ ana [ˡᵘ́ZU⁻ᵘ́] | li-kal-lim la ˡᵘ́ZU⁻ᵘ́ nu [im-mar ik-kib] | ᵈa-⌈nù⌉ ᵈEN.LÍL ⌈ù⌉ [ᵈé-a DINGIR.MEŠ GAL.MEŠ] (Otto Neugebauer, *Astronomical Cuneiform Texts, Volumes I–III* (Berlin: Springer, 1955), no. 135U rev. 12–16; cf. Kathryn Stevens, "Secrets in the Library: Protected Knowledge and Professional Identity in Late Babylonian Uruk," *Iraq* 75 (2013): 211–53, at 252 no. 45.

12 This article arises from the UK AHRC-funded research project *The Geography of Knowledge in Assyria and Babylonia* (AH/E509258/1), which I ran at the University of Cambridge, 2007–12 (http://oracc .org/cams/gkab). The project website includes online editions of the scholarly

1.2 Old and New Approaches to the Topic

Assyriologists have sought for a long time to identify the textual features and genres of cuneiform scholarship that attracted protective formulae. For much of the twentieth century, the study of Mesopotamian intellectual history was tightly focused on the production of text editions in order to (re)construct the textual evidence base. It was therefore natural to assume that ancient motivations for protecting works of cuneiform scholarship lay in the texts themselves: that they represented a body of *Geheimwissen*, "secret knowledge", that had to be divinely protected from outsiders at all costs. Concluding an extensive survey of earlier work in the field, as part of his own investigations into the phenomenon, Alan Lenzi admitted defeat.[13] It was "a dead-end," he argued, to even ask why particular compositions or textual genres were marked as secret knowledge, as this label was "applied inconsistently" to works of cuneiform scholarship. One amongst many otherwise identical manuscripts of a particular composition might invoke divine protection, though the others do not. One chapter of a scholarly work might be marked as "secret," the others not. The very parameters of cuneiform esotericism were apparently so esoteric as to be utterly inscrutable, even to the modern ranks of the "learned."

More recently, Kathryn Stevens persuasively demonstrated that earlier generations of historians have been missing a trick.[14] Rather than treating *Geheimwissen* as a property of the texts themselves, we should see the secrecy label as just one of several types of protective strategies. Such formulations, she argues, were an expression of "clearly articulated relationships between the professional specialism(s) of the individual scholar and the texts he sought to protect."[15] Her case study was the small, close-knit intellectual community of the Babylonian city of Uruk in the fifth to third centuries BCE, where Anu-aba-uter and Anu-belšunu lived and worked. Their circle comprised men from just three or four extended families, each named after an eponymous ancestor, and each specialising in one or two venerable scholarly professions.

Descendants of Sin-leqe-unninni, such as Anu-belšunu, called themselves *kalûs*, "lamenters," specialists in soothing the hearts of angered gods though prayer, ritual and lamentation. Members of the Šangu-Ninurta, Ekur-zakir and Hunzu clans self-identified as *āšipus*, often translated rather awkwardly into

---

tablets from Huzirina, Kalhu and Uruk discussed here. I am most grateful to Kathryn Stevens for her constructive and perspicacious comments on the final draft.
13 Lenzi, *Secrecy and the Gods*, 214.
14 Stevens, "Secrets in the Library."
15 Stevens, "Secrets in the Library," 231.

English as "exorcist" or "incantation-priest" but whose main role was to heal their clients through physical therapy or ritual reconciliation with the divine. A few of the more numerate men in each family also trained as *ṭupšar Enūma Anu Ellil*, literally "scribes of the celestial omen series 'When the gods Anu and Ellil,'" usually rendered as "astrologer," By this late period, short-term divination through observing the moon and planets was obsolete, as the precise movements of the major heavenly bodies could be determined mathematically. Instead the Hellenistic *ṭupšar Enūma Anu Ellil* developed increasingly sophisticated methods for predicting lunar and planetary motion, testing them against night-time observations. They also drew up horoscopes for private clientele. Each generation taught members of the next, usually sons and nephews, but also youngsters of the other families, as well as members of the elite Ahu'tu clan, which produced several of Uruk's city governors. All of these men, and many other members of their extended families, also drew income and social status from prebends, or rights to temple income, in return for a few days of ritual duty a year.[16]

Stevens showed that in Late Babylonian Uruk each composer or copyist of cuneiform scholarship chose whether or not to invoke protective formulae in the colophons of the texts they wrote.[17] Men with the title *āšipu* or *kalû* were most likely to protect works most closely associated with their respective professional specialisms but not to bother protecting those that were intellectually interesting but not closely tied to personal professional identity. This was not a hard and fast rule, but clear trends were visible. In the temple the primary duty of *kalûs* such as Anu-belšunu, for instance, was to soothe and sympathise with the gods in their times of distress – one of those times being during a lunar eclipse. Knowing precisely when such eclipses would occur enabled them to perform their lamentation rituals with ultimate efficacy. Eclipse tables were thus at the intellectual heart of the *kalûs*' cultic role, overseen by the sky-god Anu with the great gods Ellil and Ea on either side of him. It made complete sense for young Anu-aba-uter to invoke their protection as he calculated potential times of divine upset.

Yet even Stevens's major breakthrough does not give a complete answer. It does not explain why some individuals and communities did *not* invoke secrecy

---

16 On the principles of Babylonian prebendary priesthood see Caroline Waerzeggers, "The Babylonian Priesthood in the Long Sixth Century BC," *Bulletin of the Institute of Classical Studies* 54 (2011): 59–70. The literature on cuneiform scholarship in Late Babylonian Uruk is extensive; see, with many further references, most recently Eleanor Robson "The Socio-economics of Cuneiform Scholarship after the 'End of Archives': Views from Borsippa and Uruk," in *At the Dawn of History: Ancient Near Eastern Studies in Honour of J. N. Postgate*, ed. Yagmur Heffron, Adam Stone, and Martin Worthington (Winona Lake: Eisenbrauns, 2017), 455–70.
17 Stevens, "Secrets in the Library."

clauses or protective formulae even on their most precious scholarship, and nor does it address the question of who the supposed perpetrators might have been. In what follows I take Stevens's model as a starting point to consider which scholarly groups felt their written knowledge to be most and least at risk, and from whom. I shall also draw on recent work on the social geographies of cuneiform scholarship, as the spread and status of high cuneiform culture diminished over the course of the first millennium BCE.[18]

As I shall argue here, the overarching threat was not from below, via the widespread adoption of alphabetic literacy, but rather from above. In the mid-first millennium cuneiform scholarship underwent two major "survival bottlenecks," to borrow a phrase from conservation biology: near-catastrophic events that threaten a population's survival, through significantly reducing its size and diversity. The first of those began with Assyrian king Ashurbnanipal's large scale appropriation of cuneiform scholarship, peaking after the civil war against his brother Šamaš-šumu-ukin in 648 BCE and culminating in the collapse of the Assyrian Empire three decades later. The second comprised a systemic attack on Babylonian temple communities as sources of political dissent and rebellion, instigated by the Achaemenid king Darius in 521 BCE and culminating in a thorough purge by his son Xerxes II in 484 BCE. Although cuneiform scholarship survived both bottlenecks, it was badly compromised each time, and had to adapt to significantly less favourable circumstances thereafter. The motivations and strategies employed for protecting learned writings can only be fully understood, I argue, in this wider political context.

The rest of this paper is thus in three parts. I shall begin by considering four communities of textual production in eighth and seventh-century Assyria, which each shared and protected their knowledge to different degrees. In the middle section I expand on the Assyrian and Achaemenid royal actions that resulted in survival bottlenecks for cuneiform scholarship and consider their long-term repercussions. In

---

[18] Eleanor Robson, "Empirical Scholarship in the Neo-Assyrian Court," in *The Empirical Dimension of Ancient Near Eastern Studies*, ed. Gebhardt Selz and Klaus Wagensonner (Vienna: LIT, 2011), 603–30; eadem, "Reading the Libraries of Assyria and Babylonia," in *Ancient Libraries*, ed. Jason König, Katerina Oikonomopoulos, and Greg Woolf (Cambridge: Cambridge University Press, 2013), 38–56; eadem, "Tracing Networks of Cuneiform Scholarship with Oracc, GKAB and Google Earth," in *Archaeologies of Text: Archaeology, Technology and Ethics*, ed. Matthew Rutz and Morag Kersel (Oxford: Oxbow Books, 2014), 142–63; eadem, "The Socio-economics of Cuneiform Scholarship"; eadem, *Ancient Knowledge Networks: A Social Geography of Cuneiform Scholarship in the First Millennium BC* (forthcoming); Eleanor Robson and Kathryn Stevens, "Scholarly Tablet Collections in First-Millennium Assyria and Babylonia," in *The Earliest Libraries: Library Tradition in the Ancient Near East*, ed. Gojko Barjamovic and Kim Ryholt (Oxford: Oxford University Press, forthcoming).

the final part before the conclusion I look at the strategies of secrecy versus sharing in Late Babylonian contexts. I revisit Stevens' work on late Achaemenid and Hellenistic Uruk, situating it in this wider context. Lastly I come to the very end of the cuneiform tradition in c. 100 BCE. As the very last known practitioners of their disciplines, what motivations did the scholars of Parthian Babylon have to share and protect scholarly knowledge that was widely considered obsolete?

1.3 Sharing and Protecting Scholarship in the Assyrian Empire

Two seventh-century scholarly communities exhibit the classic model of sharing and protecting knowledge in cuneiform culture, around a so-called "distributed library."[19] In the ancient city of Assur, cultural heart of the Assyrian empire and close to the seat of power, the Baba-šumu-ibni family worked as *āšipu*-healers, affiliated to the god Aššur's temple Ešarra. When Assur fell to the invading Medes and Babylonians in 614 BCE, the family left behind some 600 scholarly tablets in their city-centre house, about of a quarter of which have colophons showing that they were written over four generations by their own family members and at least thirteen unrelated apprentices.[20] Nearly three-quarters of their writings relate somehow to their profession: medical recipes, rituals and incantations; but they also include temple ceremonies, hymns and prayers, and a small collection of bilingual "lexical lists" which explicated the complexities of cuneiform script and the subtle relationships between Sumerian and Akkadian vocabulary.[21] Meanwhile, some 430 km to the northwest in the politically important province of Harran, several generations of the Nur-Šamaš family of *šangû*-priests ran a scribal school for the sons of mid-ranking imperial officials.[22] It operated in the

---

[19] Robson and Stevens, "Scholarly Tablet Collections in First-Millennium Assyria and Babylonia."

[20] Stefan M. Maul, "Die Tontafelbibliothek aus dem sogenannten »Haus des Beschwörungspriesters,«" *Assur-Forschungen: Arbeiten aus der Forschungsstelle »Edition Literarische Keilschrifttexte aus Assur« der Heidelberger Akademie der Wissenschaften*, ed. Stefan M. Maul and Nils P. Heeßel (Wiesbaden: Harrassowitz, 2010), 189–228.

[21] The research project Edition literarischer Texte aus Assur, led by Professor Stefan Maul at the University of Heidelberg, is systematically publishing the scholarly texts from this house and elsewhere in Assur (http://www.haw.uni-heidelberg.de/forschung/forschungsstellen/keilschrift/index.de.html, accessed 9 September 2016).

[22] Robson, "Tracing Networks of Cuneiform Scholarship," 152–53. The Huzirina tablets were published in scale drawings by Gurney and Finkelstein, *The Sultantepe Tablets*; Oliver R. Gurney

small town of Huzirina for at least a hundred years, until it too was abandoned at the very end of empire in the late seventh century BCE. When the last occupants left the building they carefully hid away nearly 400 tablets in the hope that they would one day return for them. This collection includes a similar proportion of incantations and rituals to that of the Assur *āšipus*, but a smaller quantity of medical recipes and a relatively larger number of hymns, omen collections and literary works. About sixty of the tablets have surviving colophons, written by the Nur-Šamaš men and at least twenty different "apprentices," *šamallû*.

Although hundreds of kilometres apart and serving very different scholarly communities – professional urban healers, imperial administrators aspiring to a cultured education – these two families shared a common attitude to knowledge and who could access it. On the one hand they protected their tablets against theft and loss, but they also made copies for others to read. For instance, Nabu-rehtu-uṣur's colophon to The Poor Man of Nippur, already quoted above, says in full:

> Written and checked [(from an original)]. [Handiwork of] Nabu-rehtu-uṣur, scribal apprentice, pupil of Nabu-ahu-iddina, eunuch, for the viewing of Qurdi-Nergal.
>
> Whoever takes away (this tablet), may Ea take him away! At the command of Nabu, who lives in Ezida, may he have no descendants, no offspring.
>
> In the month Addaru (Month XII), on the 21st day, eponymate of Hanani, the provincial governor of Til-Barsip (701 BCE).
>
> Do not take away the tablets! Do not disperse the collection! Taboo of the god Ea, king of the Abyss.[23]

Likewise, one of the Assur *āšipus* writes the following at the end of a ritual to dispel the evil of a dog which has misbehaved in his client's house:

> Written and checked according to the wording of its original. Tablet of Nabu-bessunu, *āšipu* of Aššur's temple, son of Baba-šumu-ibni the chief *āšipu* of the Ešarra temple.
>
> Whoever takes away this tablet, may the god Šamaš take away his eyes![24]

---

and and Peter Hulin, *The Sultantepe Tablets, Volume II* (London: British Institute of Archaeology at Ankara, 1964). For an up-to-date catalogue, bibliography and online edition see http://oracc.org/cams/gkab.

23 ša IR ^{d}60 lit-bal-šú | ina qí-bit ^{d}MUATI a-šib É.ZI.DA | a-a GÁL$^{-ši}$ NUNUZ-šú na-an-nab-šú ṭup-pi la ⌈ta-ta⌉-bil | ^{im}GÚ.[LÁ] la ta-par-ra-ru | [ik]-⌈kib⌉ ^{d}60 LUGAL ABZU (Gurney and Finkelstein, *The Sultantepe Tablets*, no. 38 rev. ii 11–13, 16–18).

24 ina KA SUMUN.BI SAR ⌈IGI.KÁR⌉ | IM ^{md}UMBISAG$_2$-be-su-⌈nu⌉ ^{lú}MAŠ.MAŠ É ^{d}[aš-šur] | PEŠ ^{md}ba-ba$_6$-[MU]-DÙ $^{⌈lú⌉}$ZABAR.DAB.⌈BA⌉ É.ŠÁR.RA | IR IM BI ^{d}UTU IGI.MIN.MEŠ-šú lit-bal (Hunger, *Babylonische und Assyrische Kolophone*, no. 193 rev. 22–27; Stefan M. Maul, *Zukunftsbewältigung:*

Colophons such as these reveal, first, that tablets were copied from other manuscripts, which must have moved from place to place and from person to person in order for this to happen. In the examples quoted above, the details of the original are lost or considered unimportant, but both collections include copies made from manuscripts from Babylon and from the goddess Gula's temple in Assur. There are also manuscripts originating from Nineveh and Uruk amongst the Assur *āšipus*'s tablets.[25] It was perhaps good manners to acknowledge one's sources, especially if copying from an individual or institution; and it also helped to keep track of the origins of variant recensions; but it could also be a matter of prestige to have access to material from glamorous, far-away cities or powerful temples.

Second, tablets could be produced precisely in order for others to read them. In Huzirina the recipient was most often Qurdi-Nergal of the Nur-Šamaš family, as in the example above, but tablets could also be intended for more than one person[26]:

> Writer: Nabu-eṭiranni. In Kislimu (Month IX), on the 26th day, eponymate of Nergal-šarru-uṣur, chief cupbearer (678 BCE).
>
> For the viewing of Bel-ah-iddin, the *šangû*-priest; [for the viewing of ...]-Ninurta; [for] the viewing of [...]-...-uṣur, the novice; for the viewing of Rimut-ilani, the junior *asû*-healer; for the viewing of Zer-ukin, the junior scribal apprentice: it has been quickly excerpted for their viewing.[27]

Eine Untersuchung altorientalischen Denkens anhand der babylonisch-assyrischen Löserituale (Namburbi) (Mainz: von Zabern, 1994), 12–23; idem, "Die Tontafelbibliothek," 195).
25 Babylonian originals at Huzirina: Gurney and Hulin, *The Sultantepe Tablets*, nos. 136, 232, 323; at Assur: Hunger, *Babylonische und Assyrische Kolophone*, no. 203I; manuscripts from Gula's temple in Assur at Huzirina: Gurney and Finkelstein, *The Sultantepe Tablets*, no. 73; at Assur: Hunger, *Babylonische und Assyrische Kolophone*, nos. 199D, 202A, 203K; from Nineveh and from Uruk at Assur: Hunger, *Babylonische und Assyrische Kolophone*, nos. 203B, 211, 212A.
26 Other tablets for Qurdi-Nergal's viewing: Gurney and Hulin, *The Sultantepe Tablets*, nos. 161, 172.
27 šà-ṭír ᵈMUATI- KAR⁻ⁱʳ-an-ni | ina ⁱᵗⁱGAN U₄ 26-KÁM | lim-mu | ᵐᵈU.GUR-MAN-PAB | ˡúGAL.KAŠ. LUL | a-na IGI.DU₈.A | ᵐᵈEN-PAP-AŠ | ˡúÉ.BAR | [...]-⁻ᵈ⁻IGI.DU | [...] ⸢IGI⸣.DU₈.A | [...] x-x-x-ŠEŠ | a-⸢ga-aš⸣-gu-ú | [a]-⸢na⸣ IGI.DU₈.A | ⸢ᵐ⸣ri-mut-DINGIR-MEŠ⁻ⁿⁱ | ˡúA.ZU ṣe-eh-ri | a-na IGI.DU₈.A ᵐNU-MUN-GUB | ˡúšam-lù-ú | ⸢ṣe⸣-eh-ri | [a]-⸢na⸣ IGI.DU₈-šú-nu | [ha]-⸢an⸣-ṭiš ZI⁻ⁱʰ (Gurney and Hulin, *The Sultantepe Tablets*, no. 301 rev. ii 11'–iii 12'; Alasdair Livingstone, "On the Organized Release of Doves to Secure Compliance of a Higher Authority," in *Wisdom, Gods and Literature: Studies in Assyriology in Honour of W.G. Lambert*, ed. Andrew R. George and Irving L. Finkel [Winona Lake: Eisenbrauns, 2000], 375–88: source GG).

Amongst the Assur *āšipus*, however it was much more usual to "quickly excerpt in order to grasp what to do," *ana ṣabāt epēši hanṭiš nasāhu*.²⁸ As Stefan Maul pointed out in his discussion of the Baba-šumu-ibni family, the phrase *hanṭiš* (or *zamar*) *nasha*, "quickly excerpted," indicates that the copy was made under time pressure for use in therapeutic practice, presumably from an original that had been borrowed and needed to be returned, or which had been copied in situ elsewhere.²⁹

In other words, tablets *did* circulate, sometimes over long distances – it was a journey of well over 700 km upriver from Babylon to Huzirina, for instance – but under closely prescribed circumstances. Given that tablets could and should move around, it was important to regulate those movements, whether by naming the intended recipients individually, or by warning borrowers not to become thieves. Written knowledge was a scarce and precious commodity: sharing what one had, within socially acceptable parameters, was an important means of enabling access to more, owned by others. Protection clauses reminded members of the group of the social contract entailed in borrowing and copying, and the professional ostracism at stake should it be transgressed.

These markers of the "distributed library," as Robson and Stevens term it, whereby professional and scholarly knowledge circulates within a self-policing community, in both text and in memory, are not restricted to seventh-century Assyria; as we argue in that paper, they are also attested amongst the *āšipus* and *kalûs* of Late Babylonian Uruk discussed briefly above.³⁰ However, they are not universally attested, even in seventh-century Assyria. By turning our attention to communities which did not protect their writings with written admonitions, we will get a clearer sense of what this practice meant.

The Issaran-šumu-ukin and Gabbu-ilani-ereš families had produced advisors to Assyrian monarchs since at least the early ninth century BCE, when the main royal residence moved to Kalhu, some 70 km up the river Tigris from Assur. The two scholarly families made their devotional base the newly founded Ezida, temple of Nabu, god of wisdom, on the royal citadel. Here they built up a collection of scholarly writings, stored in a dedicated room immediately opposite Nabu's inner shrine. When invaders sacked the Ezida temple in 612 BCE, at least 250 tablets

---

28 E.g., a-na ṣa-bat e-pe-ši ha-an-ṭiš na-as-ha (Hunger, *Babylonische und Assyrische Kolophone*, no. 197A–E; 198A–C).
29 Maul, "Die Tontafelbibliothek," 212–13; Gurney and Finkelstein, *The Sultantepe Tablets*, nos. 4, 57 are also said to be "quickly excerpted."
30 Robson and Stevens, "Scholarly Tablet Collections in First-Millennium Assyria and Babylonia."

were still in situ, including around 30 with colophons on them.[31] They name men from several generations of the two families – although never both together – but there is very little evidence for training of apprentices, as in Huzirina and Assur.[32] As befitted royal advisors and healers, about a quarter of the families' collection comprised omens for divining the gods' intentions for the land and a further quarter consisted of medical recipes, incantations and rituals. There were also significant numbers of hymns and prayers, lexical texts, and royal inscriptions. A further 85 or more scholarly tablets that had formerly belonged to Gabbu-ilani-ereš men also made their way into the royal palace collections in Nineveh – a fact that we shall return to, and account for, in the following section.[33]

Issaran-šumu-ukin's successors fell out of royal favour in the eighth century but the descendants of Gabbu-ilani-ereš continued to serve at court until at least 650 BCE, as royal *āšipus*, *ṭupšar Enūma Anu Ellil* and even as senior scholar, *ummânu* or *rab ṭupšarrī* (literally "expert" or "chief scribe"; the two titles are synonymous). By this time the king was mostly resident in Nineveh, some 35 km upstream from Kalhu. The palace archives include about 1500 scholarly letters and reports to kings Esarhaddon and Ashurbanipal, some ten percent of which are from Gabbu-ilani-ereš men. They advise the king on state affairs through divination, look after the royal family's health, and take care of royal ritual. But the scholars were not always by his side, as chief *āšipu* Adad-šumu-uṣur explains to Esarhaddon in about 670 BCE:

> Concerning what the king, my lord, wrote to me: "Why haven't you sent an answer to (my) letter?" – I had to drive to the palace those rams that the chief cook had brought out for

---

[31] The tablets from Ezida were published as scale drawings by Donald J. Wiseman and Jeremy A. Black, *Literary Texts from the Temple of Nabû*, Cuneiform Texts from Nimrud 4 (London: British School of Archaeology in Iraq, 1996), up-to-date bibliography, catalogue and online edition at http://oracc.org/cams/gkab; recent studies by Robson, "Reading the Libraries," 45–48; eadem, "Tracing Networks of Cuneiform Scholarship," 148–51.

[32] A "junior scribal apprentice," ŠAMAN₂.LÁ TUR is mentioned on Wiseman and Black, *Literary Texts*, no. 27 rev. ii 9', on which see further below with note ; and a [...] ṣeh-ru, "junior [...]," on Wiseman and Black, *Literary Texts*, no. 220 rev. ii 4', a manuscript of the synonym list *Malku = Šarru*.

[33] For the tablets of Nabu-zuqup-kena see Stephen Lieberman, "A Mesopotamian Background for the So-called Aggadic 'Measures' of Biblical Hermeneutics?" *Hebrew Union College Annual* 58 (1987): 157–225, at 204 n. 222; Eckart Frahm, "Nabu-zuqup-kenu, das Gilgameš-Epos und der Tod Sargons II," *Journal of Cuneiform Studies* 51 (1999): 73–90, at 88; for those of his descendants see Simo Parpola, *Letters from Assyrian Scholars to the Kings Esarhaddon and Assurbanipal, Part II: Commentary and Appendices* (Kevelaer: Butzon & Bercker, 1983; repr., Winona Lake: Eisenbrauns, 2007), 450–53.

me, and the writing-board was at my house. Now then, I can look at the writing-board and extract the relevant interpretation. Concerning the ritual against earthquake [...].³⁴

As wealthy and influential courtiers, Adad-šumu-uṣur and his relatives doubtless had homes in Nineveh, but this letter suggests that not all of their scholarly collection was close to hand. It is highly likely that the Gabbu-ilani-ereš family base remained in Kalhu, along with their tablets in the temple there.

The royal scholars of Kalhu, whether descendants of Issaran-šumu-ukin or Gabbu-ilani ereš, display a very different attitude to the protection and sharing of knowledge than their less powerful contemporaries in Assur and Huzirina. They regularly acknowledge that their manuscripts are copies of earlier exemplars but *never* announce that materials have been "quickly excerpted" from them and *never* acknowledge the identity of the scribe making the copy for them.³⁵ If they borrow tablets from others, there is apparently no hurry to return them. Nor do they appear to countenance sharing or lending. Just once does Nabu-zuqup-kena describe a work as "a secret of the sages" which "the unlearned may not see."³⁶ The only people that are said to "view" tablets are the owners themselves or their sons. In 711 BCE, for instance, Nabu-zuqup-kena had at least four tablets of divination made for him, all of which end with the comment:

---

34 ša LUGAL be-lí | iš-pur-an-ni ma-a a-ta-a | GABA.RI e-gír-ti la taš-pur-ra | ina ŠÀ É.GAL a-na ⸢UDU.NÍTA⸣-MEŠ šú-nu | ša ᴸᵘGAL – MU ú-še-ṣa-an-ni | ú-se-li ᵍⁱšZU ina É šú-u | ú-ma-a an-nu-rig ᵍⁱšZU | a-mar pi-šìr-šu a-na-sa-ha | ina UGU dul-li ša ri-i-bi (Simo Parpola, *Letters from Assyrian and Babylonian Scholars*, State Archives of Assyria 10 [Helsinki: Helsinki University Press, 1993], no. 202 obv. 5–13).

35 A detailed analysis by Lieberman, "A Mesopotamian Background," 209–10 showed that Nabu-zuqup-kenu of the Gabbu-ilani-ereš family employed several different copyists.

36 ni-ṣir-ti NUN.ME là mu-du-ú là IGI⁻ᵐᵃʳ (K 170 + Rm 520 rev 9', an extract from the scholarly commentary *Inam gišhurankia*, ed. Alasdair Livingstone, *Mystical and Mythological Explanatory Works of Assyrian and Babylonian Scholars* [Oxford: Clarendon, 1986], 30–33; cf. Lenzi, *Secrecy and the Gods*, 174). Even if Hunger, *Babylonische und Assyrische Kolophone*, no. 311 and Lieberman, "A Mesopotamian Background," 205 n. 220 are right to assign the anomalous fragment K 11867 to Nabu-zuqup-kena, the phrase on the colophon of that tablet which Hunger reads as [...] ⸢GI?⸣ ZU⁻ᵃ li-ka₁₅-lim "... may show the learned," seems highly doubtful as it depends on the otherwise unattested reading ka₁₅ for the sign GAR and does not account for the damaged sign at the beginning of the phrase, whose traces do not fit the expected writing ZU⁻ᵃ for *mūdû*, "learned." In my opinion, it is more likely to be read as [...] he-⸢pa⸣-a li-šá-lim, "may he restore the breaks" (cf. the exact parallel phrase in RIMB 2.02.08.05 discussed below in note 39).

According to the words of an old wooden writing-board of Amel-Uraš-liya(?), son of Esangila-zeru-iddin the diviner. Nabu-zuqup-kena, son of Marduk-šumu-iqiša, descendant of Gabbu-ilani-ereš the chief scribe, had it written and checked it for his (own) viewing.[37]

The only surviving admonitions to future readers relate not to careful treatment of the tablet but to consideration of the quality of the copy. In 684 BCE Nabu-zuqup-kena, now an old man, explains that he produced a copy of the scholarly commentary *Inam gišhurankia*:

> For the viewing of Ištar-šumu-ereš my (grand)son. 1 1/2 years ago my vision deteriorated but I hurriedly bestirred and wrote it. The viewer should not disparage (my efforts).[38]

Similarly, another of Nabu-zuqup-kena's grandsons, Urad-Gula, explains the origins of an old dedicatory inscription he has copied:

> Written and checked according to its original. Written according to the wording of damaged tablets. Anyone who views (it) should not disparage (my efforts). (Instead), let him restore the break(s)![39]

In other words, these men are concerned more with protecting their scholarly reputations than with safeguarding the contents of their writings.

Only three tablets from the collection in the Ezida have (the remains of) divine protection formulae on them: one written by a junior scribe whose name no longer survives[40] and two by an *āšipu*-healer named Banunu:

---

37 E.g., ki-i pi-i ᵍⁱˢle-u₅-um LIBIR.RA ša ᵐLÚ-ᵈURAŠ-li-ia DUMU ᵐÉ.SAG.ÍL-MU ˡúHAL | ᵐᵈNÀ-zu-qu-up-GI.NA DUMU ᵐᵈma-ru-duk-MU-BA⁻ˢᵃ́ ˡúDUB.SAR⁻ʳⁱ | ŠÀ.BAL.BAL ᵐgab-bu-DINGIR.MEŠ⁻ⁿⁱ-KAM⁻ᵉˢ̌ ˡúGAL DUB.SAR.MEŠ | a-na ta-mar-ti-šú ú-šá-áš-ṭir-ma íb-ri (Hunger, *Babylonische und Assyrische Kolophone*, no. 297A; cf. Lieberman, "A Mesopotamian Background," 210).

38 ᵐᵈNÀ-zu-qup-GI.NA DUMU ᵐᵈAMAR.UTU-MU-BA⁻ˢᵃ́ ˡúDUB.SAR [...] | a-na ta-mar-ti ᵐᵈINNIN-MU-KAM⁻ᵉˢ̌ DUMU-ia ul-tu 1 1/2 MU.AN.NA.MEŠ di-ig-la ú-kab-bir-ma za-mar ú-ba-ah-hi-iš-ma ab-r[i? ...] | a-mi-ru la i-ṭa-ap-pil (Hunger, *Babylonische und Assyrische Kolophone*, no. 299, ed. Livingstone, *Mystical and Mythological Explanatory Works*, 29; cf. Lieberman, "A Mesopotamian Background," 213–14).

39 LIBIR.RA.BI.GIM AB.SAR BA.AN.È | i-na KA ṭup-pi GAZ.MEŠ šà-ṭir a-me-ru la i-ṭa-pil he-pa-a ⌈li⌉-šal-lim | ⌈DUB⌉ ᵐÌR-ᵈGU.LA ˡúMAŠ.MAŠ.ME.EN | ⌈DUMU⌉ ᵐᵈIŠKUR-MU-ú-ṣur ˡúšá-an-gam-ma-⌈hu⌉ ša ʳᵐAN.ŠÁR-PAP-AŠ⌉ MAN ⌈KUR⌉ aš-šurᵏⁱ (Hunger, *Babylonische und Assyrische Kolophone*, no. 498 12–17; Grant Frame, *Rulers of Babylonia: From the Second Dynasty of Isin to the End of Assyrian Domination (1157–612 BC)*, Royal Inscriptions of Mesopotamia: Babylonian Periods 2 [Toronto: University of Toronto Press, 1995], no. 2.02.08.05, ex. 01; Parpola, *Letters from Assyrian Scholars*, 453 no. 15).

40 On a tablet of the astronomical compilation *Mul-Apin*: [...] ⌈GIM⌉ SUMUN-šú SAR-ma bà-rì DUB [...] | [...] ŠAMAN₂.LÁ TUR ša IR ᵈUTU [...] | ⌈ina⌉ dan-na-⌈ni⌉ e-kim-šu⌉ "Written and checked

Written and checked according to its original. Tablet of Banunu, *āšipu*. Do not deliberately(?) remove (it). Do not disperse the collection. Taboo of the god Ea, king of the Abyss.[41]

Banunu is also known as the copyist of three further scholarly tablets found in Ezida, but never gives a father's name or any family affiliation.[42] Was he perhaps a eunuch? Although he may sometimes have worked at court,[43] he did not have the status or genealogy of the Issaran-šumu-ukin or Gabbu-ilani-ereš men, and may have thus felt the need for divine protection more keenly than they did. Nevertheless, like the Gabbu-ilani-ereš men he did not feel the need to credit his copyists.[44]

There was just one man in seventh-century Assyria who felt more confident than the royal scholars in the security of his tablets, and even less need to share them. Since at least the time of Sargon II (r. 721–705 BCE) kings had collected tablets for the palace, but his great grandson Ashurbanipal (r. 668–c.630 BCE) took that tradition to its logical extreme.[45] Neither he nor his father Esarhaddon had been first in line for the throne, and thus Ashurbanipal grew up in train-

---

like its original. Tablet of [...], junior apprentice scribe. Whoever takes it away, may Šamaš [...] remove him by force" (Wiseman and Black, *Literary Texts*, no. 27 rev. ii 8'–10').

41 On a tablet of the induction ritual for a cult statue, *Mīs Pî*: ⌈LIBIR.RA⌉.BI.GIM AB.SAR.ÀM-ma ⌈BA.AN.E₃⌉ | DUB ᵐba-nu-ni ˡúMAŠ.MAŠ ina <me>-reš-tù [...] là TÙM | IM.GÚ.LA là BAR-ár NÍG. ⌈GIG⌉ ᵈé-a LUGAL ABZU (Wiseman and Black, *Literary Texts*, no. 170 (+) 188 rev. ii 5'–7') ed. Daisuke Shibata, "A Nimrud Manuscript of the Fourth Tablet of the Series *Mīs pî*, CTN IV 170(+)188, and a *Kiutu* Incantation to the Sun God," *Iraq* 70 [2008]: 189–203; cf. Wiseman and Black, *Literary Texts*, no. 116, a collection of medical recipes and incantations against wounds, ed. Markham J. Geller, "Fragments of Magic, Medicine and Mythology from Nimrud," *Bulletin of the School of Oriental and African Studies* 63 [2000]: 331–39, at 336–339).

42 GABA.RI KÁ.DINGIR.RAᵏⁱ LIBIR.RA.BI.[GIM] | IN.SAR-ma BA.AN.⌈È⌉ | DUB ᵐba-nu-ni ˡúMAŠ. MAŠ "Manuscript of Babylon. Written and checked [like] its original. Tablet of Banunu, *āšipu*"; ⌈DUB⌉ ᵐba-nu-ni ˡúMAŠ.MAŠ "Tablet of Banunu, *āšipu*" (Wiseman and Black, *Literary Texts*, no. 61 + 62 rev. iii 5'–7'; no. 63 rev iii 28): Tablet 7 and 9 of a series of prayers to Šamaš, god of divination, ed. Wilfred G. Lambert, Babylonian Oracle Questions (Winona Lake: Eisenbrauns, 2007); LIBIR.RA.BI.GIM AB.SAR [...] | ṭup-⌈pi⌉ ᵐba-nu-ni ˡúMAŠ.MAŠ [...] "Written [and checked] like its original. Tablet of Banunu, *āšipu*" (Wiseman and Black, *Literary Texts*, no. 192 rev. ii 6"–7"), the plant list *Uruanna*.

43 See F. Mario Fales and J. Nicholas Postgate, *Imperial Administrative Records, Part II: Provincial and Military Administration*, State Archives of Assyria 11 (Helsinki: Helsinki University Press, 1995), no. 156, to which we will return shortly.

44 Wiseman and Black, *Literary Texts*, 15.

45 For instance, the cuneiform inscription on a cover of a writing-board found at Kalhu reads, É.GAL ᵐMAN-GI.NA MAN kiš-šá-ti | MAN ᵏᵘʳaš-šurᵏⁱ * U₄ AN ᵈEN.LÍL ÉŠ.GÀR | ina ᵍⁱšle-u₅-um ZÚ AM.SI ú-šá-áš-ṭir-ma | ina qé-reb É.GAL-šú ina ⁱʳⁱBÀD-MAN-GIN ú-kin, "Palace of Sargon, king of the world, king of Assyria. He had the series *Enūma Anu Ellil* written on a writing-board of elephant-ivory and deposited it in his palace at Dur-Šarruken" (ND 3557; Donald J. Wiseman, "Assyrian Writing Boards," *Iraq* 17 [1955]: 3–13, at 7).

ing for the priesthood not for kingship.⁴⁶ Ashurbanipal's literacy and fascination with cuneiform scholarship has been extensively studied and discussed.⁴⁷ He also made use of his knowledge in the practice of kingship, insisting that diviners send him their observations so that he could check their interpretations and advice against the written tradition.⁴⁸ There is no doubt that, building on already substantial royal collections, he amassed a vast "library" of tablets and writing boards for his own private use, especially focused on divination, the extant remains of which comprise around 27,000 tablets and fragments now held in the British Museum.⁴⁹

As the royal citadel of Nineveh was dug primarily by the first generations of Victorian explorers, long before the advent of stratigraphic archaeology, it is now almost impossible to reconstruct exactly what was found where.⁵⁰ In broad outline, however, scholarly tablets were kept in at least two palaces and one or more temples on the citadel, all of which were ransacked during the final destruction of Nineveh in 612 BCE. This means that even if the find contexts had been recorded to current standards, they would show only the tablets' final resting places after the looting, not their normal storage arrangements.

Nevertheless, the tablets themselves shed a good deal of light on the circumstances of their production and intended use. Let us start with the colophons that Ashurbanipal had inscribed on "almost every tablet of importance in the … collection."⁵¹ Stephen Lieberman divides them into three broad categories.⁵²

---

46 Alasdair Livingstone, "Ashurbanipal: Literate or Not?" *Zeitschrift für Assyriologie* 97 (2007): 98–118, at 99.
47 E.g., Pierre Villard, "L'education d'Assurbanipal," *Ktema* 22 (1997): 135–49; Livingstone, "Ashurbanipal"; Eckart Frahm, "Keeping Company with Men of Learning: The King as Scholar," in *The Oxford Handbook of Cuneiform Culture*, ed. Karen Radner and Eleanor Robson (Oxford: Oxford University Press, 2011), 508–33.
48 Robson, "The Production and Dissemination."
49 In addition to some 5000 letters and legal documents, now published in the *State Archive of Assyria* series and at http://oracc.org/saao/. Data from The British Museum's *Ashurbanipal Library Project*, headed by Jonathan Taylor, http://oracc.org/asbp/corpus/, accessed 10 August 2016. For a convenient recent overview, with references to further literature, see Robson, "Reading the Libraries," 41–45.
50 Julian E. Reade, "Ninive (Nineveh)," in *Reallexikon der Assyriologie und Vorderasiatischen Archäologie, Vol. 9*, ed. Dietz O. Edzard (Berlin: De Gruyter, 2001), 388–433, at 421–27.
51 Carl Bezold, *Catalogue of the Cuneiform Tablets in the Kouyunjik Collection of the British Museum, Volume 5* (London: The British Museum, 1899), xiii.
52 Stephen Lieberman, "Canonical and Official Cuneiform Texts: Towards an Understanding of Assurbanipal's Personal Tablet Collection," in *Lingering over Words: Studies in Ancient Near Eastern Literature in Honor of William L. Moran*, ed. Tzvi Abusch, John Huehnergard, and Piotr Steinkeller (Atlanta: Scholars Press, 1990), 305–36.

First, there are a few surviving witnesses to Ashurbanipal's early career, which end in the "prince" (*rūbû*) making elaborate prayerful dedications to Nabu, god of wisdom, for deposit in his temple on the Nineveh citadel.[53] These are likely to have been written by Ashurbanipal himself. Second, the large majority of scholarly tablets, produced by chancery scribes, are stamped, inscribed or painted with a simple property mark, "Palace of Ashurbanipal, king of the world, king of the land of Ashur."[54] Third, a smaller number finish with more elaborate colophons claiming that the king himself wrote, checked and deposited the tablet in the palace *ana tāmartišu* "for his (own) viewing" and similar phrases.[55] For instance:

> Tablet of Ashurbanipal, great king, strong king, king of the world, king of the land of Ashur, beloved of the great gods, to whom the gods Šamaš and Adad taught broad wisdom, who has learned and internalised divination, the secret of heaven and earth, the wisdom of Šamaš and Adad. He wrote, inspected, and checked this tablet and deposited it in his palace.[56]

Here, the "secret" is the *practice* of divination which Ashurbanipal is privy to, not the tablet himself: it is a claim about his learnedness, not a protective admonition about the tablet. Ashurbanipal had no need to invoke protective measures, for his collection was stored in the high-security environment of the royal palace where no theft was possible.[57] More than that, at one level he seems not to have acknowledged the separate existence of the scholars around him who might have wanted access to his collection.

As this colophon shows, Ashurbanipal often presented himself as a copyist of scholarship. However, as the text itself – a chapter from the sacrificial divination series *Bārûtu* – is written in the same elegant, anonymous chancery hand of all Assyrian royal output, it is highly unlikely that Ashurbanipal physically wrote it or any of the scholarly tablets produced in his name once he was king.

---

[53] Hunger, *Babylonische und Assyrische Kolophone*, nos. 328, 338, 339. For a more detailed discussion of Ashurbanipal's tablet collections, see Robson, *Ancient Knowledge Networks*, chapter 3.

[54] É.GAL ^{md}aš-šur-DÙ-IBILA LUGAL ŠÚ LUGAL ^{d}aš-šur^{ki} (Hunger, *Babylonische und Assyrische Kolophone*, no. 317).

[55] E.g., Hunger, *Babylonische und Assyrische Kolophone*, nos. 318–19, 323–25.

[56] ṭup-pu ^{m}AN.ŠÁR-DÙ-IBILA LUGAL GAL$^{-ú}$ LUGAL dan-nu LUGAL ŠÚ LUGAL KUR AN.ŠÁR^{ki} | na-ram DINGIR.MEŠ GAL.MEŠ šá ^{d}UTU u ^{d}IŠKUR šá GEŠTU.MIN DAGAL-tu$_4$ ú-šá-hi-zu-šú-ma | NAM.AZU AD.HAL AN^{-e} u KI^{-ti} né-me-qí ^{d}UTU u ^{d}IŠKUR i-hu-zu-ma | uš-ta-bi-lu ka-ras-su ṭup-pu UR$_5^{-tú}$ iš-ṭur is-niq ib-re-e-ma ina qé-reb É.GAL-šú ú-kin (Hunger, *Babylonische und Assyrische Kolophone*, no. 325).

[57] On Assyrian palace security see Karen Radner, "Gatekeepers and Lock Masters: The Control of Access in Assyrian Palaces," in *Your Praise is Sweet: A Memorial Volume for Jeremy Black from Students, Colleagues and Friends*, ed. Heather D. Baker, Eleanor Robson, and Gábor Zólyomi (London: British Institute for the Study of Iraq, 2010), 269–80.

Yet supposedly no-one else was involved in their production. Nor was anyone else meant to read them. Not a single one of Ashurbanipal's tablets carries a date of production, and not a single one bears any sort of protective formula or warning to future readers. These were the king's own tablets and no-one else at all was to share them.

Nor was anyone else to be attributed with prior knowledge. While several of Ashurbanipal's colophon types note that they have been copied from earlier sources, they never give the sort of precise information that we have seen was favoured by all scholarly groups we have looked at so far. Instead we find vague statements such as "according to the wording of original tablets (and writing-boards) from the land of Aššur and the land of Sumer and Akkad."[58] The whole of Assyria and Babylonia were at the king's intellectual disposal, in other words: no-one community or individual should be credited with particular knowledge, which now all belonged to the crown.

1.4 Destruction Events as Survival Bottlenecks for Cuneiform Scholarship

Ashurbanipal's singularly solipsistic view of himself as sole scholar was not intrinsically catastrophic for cuneiform scholarship outside the palace in Nineveh. So far as we can tell, the urban scholarly communities of Huzirina and Assur continued relatively unaffected by his actions: both the Nur-Šamaš and the Baba-šumu-ibni families continued to produce scholarly tablets, and therefore also to attract apprentices and clients, until the 610s BCE.[59] It was only then that

---

58 Hunger, *Babylonische und Assyrische Kolophone*, nos. 318, 328, 336. Rykle Borger, "Bemerkungen zu den akkadischen Kolophonen," *Welt des Orients* 5 (1969–70): 165–71, at 168, notes just one exception: four tablets of the ritual purification ritual *Bīt Rimki* were copied ki-i KA ^{giš}le-u$_5$-um/^{giš}ZU GABA.RI KÁ.DINGIR.RA^{ki}, "according to the wording of original writing-boards from Babylon." Perhaps in this case it was important to show that the ritual was steeped in genuine Babylonian tradition.

59 The latest dated tablet from the Huzirina cache is Gurney and Hulin, *The Sultantepe Tablets*, no. 300 (ed. Markham J. Geller, "Incipits and Rubrics," in *Wisdom, Gods and Literature: Studies in Assyriology in Honour of W.G. Lambert*, ed. Andrew R. George and Irving L. Finkel [Winona Lake: Eisenbrauns, 2000], 225–58), copied by a son of Nabu-zer-kitti-lešir of the Nur-Šamaš family, dated to the eponymate of Bel-ahhu-uṣur, either 621 BCE (Julian E. Reade, "Assyrian Eponyms, Kings and Pretenders, 648–605," *Orientalia* 67 [1998]: 255–65) or 616 BCE ("Sequence of Post-canonical Eponyms," in *The Prosopography of the Neo-Assyrian Empire, Volume 1/I: A*, ed. Karen Radner [Helsinki: The Neo-Assyrian Text Corpus Project, 1998], xviii–xx). The Baba-šumu-ibni

the full consequences of the king's actions, entwined with the devastating war he waged in Babylonia, were realised.

Because Ashurbanipal's grandiose project for erasing the geography of cuneiform scholarship was never completed, it has left clear evidence behind. First, there are the raw materials – other people's tablets – that were still present in the royal palaces at their destruction. Second, there is documentary evidence of the editorial process, which involved coercion as well as compliance. Third, as we shall see in the final section, Ashurbanipal's actions remained in Babylonian cultural memory for over half a millennium after his death and the fall of Assyria itself.

Even – or perhaps especially – the scholarly families closest to Ashurbanipal were subject to his acquisitive passions. We have already seen that 85 or so of the scholarly tablets written or owned by Gabbu-ilani-ereš men were found not in Nabu's temple in Kalhu but on the royal citadel in Nineveh – even though many of them explicitly say that they were written in Kalhu. Lieberman states that there is "no reason to assume that they were part of the king's library" but instead suggests that they remained in the family's possession, implying that they had a residence on the royal citadel (for which there is no archaeological evidence one way or another).[60] However, he overlooks an important piece of evidence in the form of an inventory, now in three non-joining fragments, from the royal citadel in Nineveh.[61] It originally comprised a six-column list of scholarly works that (formerly?) belonged to named individuals, including a man called Aplaya (who can be identified as a *ṭupšar Enūma Anu Ellil* from the Babylonian city of Borsippa)[62] and Esarhaddon's chief *āšipu* Adad-šumu-uṣur, whom we have already met above.

---

collection contains several works mentioning the name of king Sin-šarru-iškun (r. 623–612 BCE) (Maul, "Die Tontafelbibliothek," 204).

60 Stephen Lieberman, "Canonical and Official Cuneiform Texts: Towards an Understanding of Assurbanipal's Personal Tablet Collection," in *Lingering over Words: Studies in Ancient Near Eastern Literature in Honor of William L. Moran*, ed. Tzvi Abusch, John Huehnergard, and Piotr Steinkeller (Atlanta: Scholars Press, 1990), 305–36. At least one piece of a tablet excavated from the Kalhu Ezida in the 1950s joins another supposedly found in Nineveh by the Victorian explorers, however (Wiseman and Black, *Literary Texts*, 33 no. 229).

61 K 11922+ (Wilfred G. Lambert, "A Late Assyrian Catalogue of Literary and Scholarly Texts," in *Kramer Anniversary Volume: Cuneiform Studies in Honor of Samuel Noah Kramer*, ed. Barry L. Eichler, Jane W. Heimerdinger, and Åke W. Sjöberg [Kevelaer: Butzon & Bercker; Neukirchen-Vluyn: Neukirchener, 1976], 313–18); online edition at http://oracc.org/cams/misc/P399525

62 He wrote at least 13 divinatory reports to king Esarhaddon in the 670s BC (Hermann Hunger, *Astrological Reports to Assyrian Kings*, State Archives of Assyria 8 [Helsinki: Helsinki University

Adad-šumu-uṣur is named in the first surviving line of a piece that belonged to the bottom of the tablet. Scribal convention dictated that this name marked the end of the list of items relating to him. Because we cannot reconstruct the exact spatial relationship between the three fragments it may be that none of the compositions listed on the other two were his. But if we assume that Adad-šumu-uṣur's tablets were listed immediately below Aplaya's then they included full sets of the celestial and terrestrial omen series *Enūma Anu Ellil* and *Šumma Ālu*, "including non-canonical tablets, word-commentaries and expositions"; five classic lexical lists; the dream omen series *Zīqīqu* and the cultic topography *Tintir* = Babylon; at least eight literary works including the Babylonian Epic of Creation *Enūma Eliš* and the epics of Gilgamesh and Etana; and presumably other works on now-missing pieces of the tablet.

This list fits well with Lieberman's characterisation of Adad-šumu-uṣur's father Nabu-zuqup-kena's tablets, written in Kalhu but found in Nineveh: mostly *Enūma Anu Ellil*, as well as commentaries on it and other works about the celestial bodies; *Šumma Ālu* and other omen collections; prayers and rituals; and Tablet XII of The Epic of Gilgamesh.[63] Given the fragmentary nature of the inventory, and the fact that Nabu-zuqup-kena's ownership of tablets can only be ascertained by surviving colophons, this is an impressive overlap. Perhaps they entered the palace collection when Adad-šumu-uṣur died; perhaps he donated them himself.

Either way, such accession was part of a larger pattern of royal tablet acquisition, both voluntary and coerced, from within the king's inner circle and beyond.[64] Most famously, huge numbers of scholarly tablets and writing boards arrived in Nineveh from Babylonia after Assurbanipal's defeat of a major insurrection there in 648 BCE, led by his brother Šamaš-šumu-ukin. Seven more inventories, just as fragmentary as the one just discussed, catalogue incoming compositions, grouped, as before, by prior owner and original location.[65] About a seventh of the scholarly tablets found on the royal citadel are in Babylonian, as opposed to Assyrian handwriting, and most concern divination, Ashurbanipal's favourite

---

Press, 1992], nos. 356–68) and a letter to the queen mother (Parpola *Letters from Assyrian and Babylonian Scholars*, no. 154)

63 Lieberman, "A Mesopotamian Background," 206–8.

64 Eleanor Robson, "The Clay Tablet Book in Sumer, Assyria and Babylonia," in *A Companion to the History of the Book*, ed. Simon Eliot and Jonathan Rose (Oxford: Blackwell, 2010), 67–83; eadem, *Ancient Knowledge Networks*, chapter 3.

65 F. Mario Fales and J. Nicholas Postgate, *Imperial Administrative Records, Part I: Palace and Temple Administration*, State Archives of Assyria 7 (Helsinki: Helsinki University Press, 1992), nos. 49–56; Simo Parpola, "Assyrian Library Records," *Journal of Near Eastern Studies* 42 (1983): 1–29.

subject.⁶⁶ In this way, urban communities throughout northern Babylonia lost their writings to the king.

But, as we saw in the previous section, those origins, so carefully documented by palace administrators, had to be erased and local variation homogenised before they were fit for royal consumption. Akkullanu, *šangû*-priest of the god Aššur's temple in Assur, oversaw the production of scholarly tablets for a king, either Esarhaddon or Assurbanipal, and discussed editorial matters with him.⁶⁷ Perhaps the originals that his team copied were from his temple, the most important in the land. A short report from immediately after the war documents local men overseeing captive Babylonians inside the Assyrian palaces, copying scholarly works or reproducing them from memory. For instance, a man named Banunu (who may be the *āšipu* from Kalhu discussed above), is said to be supervising the son of the city-governor of Nippur who "has completed the series (*Enūma Anu Ellil*) and has been put in irons".⁶⁸ It was surely these men – Akkullanu, Banunu and their charges – who transformed and homogenised the many local knowledge traditions, as represented by the incoming tablets, into a uniform, timeless, geographically neutral body of learning for the king.

Ashurbanipal's grandiose editorial scheme was never completed. The four-year war against Babylonia had been vastly expensive and produced none of the usual haul of booty. Cuneiform scholarship may have had huge cultural value but it did not pay for the upkeep of the empire. Textual production petered out, the archives fell silent, the scholars disappeared from court in the course of the 640s BCE.⁶⁹ Just a few decades later, terminally weakened by Ashurbanipal's rule, the Assyrian empire finally collapsed under the weight of another drive for Babylonian independence, this time fought with the aid of Median and other allies. Assur fell in 614 BCE, Nineveh and Kalhu in 612, while the Nur-Šamaš family probably

---

66 Jeanette C. Fincke, "The Babylonian Texts of Nineveh: Report on the British Museum's Ashurbanipal Library Project," *Archiv für Orientforschung* 50 (2003/04): 111–49.
67 Parpola, *Letters from Assyrian and Babylonian Scholars*, nos. 101–3. Akkullanu's celestial omen reports for royal clients cover the period 676–650 BCE (Hunger, *Astrological Reports*, nos. 100–112; Parpola, *Letters from Assyrian and Babylonian Scholars*, nos. 84–108, 232; Stephen W. Cole and Piotr Machinist, *Letters from Assyrian and Babylonian Priests to Kings Esarhaddon and Assurbanipal*, State Archives of Assyria 13 [Helsinki: Helsinki University Press, 1998], no. 16).
68 The full passage reads: ᵐᵈMAŠ-ŠU DUMU LÚ.GÚ.EN.NA | ÉŠ.GÀR ug-da-mir | si-par-ri AN.BAR šá-kin | ina É re-du-te | ina IGI ᵐba-a-nu-ni pa-aq-qid | dul-lu ina ŠU.MIN-šú la-áš-šú "Ninurta-gimilli, the son of the *šandabakku*, has completed the Series and has been put in irons. He is assigned to Banunu in the Succession Palace but there is no work for him at present" (Fales and Postgate, *Imperial Administrative Records, Part II*, no. 156 obv. 8–13).
69 See Robson, *Ancient Knowledge Networks*, chapters 2–3 for more details.

abandoned their house in Huzirina when nearby Harran – now the last bastion of the empire – was besieged in 610 or 605 BCE.

The sacking of all of these cities and towns entailed the destruction, abandonment and eventual collapse of the buildings in which scholarly tablets were kept – from palaces and temples to family houses. All of the thousands of Assyrian tablets found in modernity by archaeologists were, de facto, tablets taken out of ancient circulation. What we can read today is precisely what was no longer accessible to read for later generations in antiquity. All that editorial work in the royal palace at Nineveh, which resulted in the recensions that modern Assyriology takes as the starting point for textual reconstruction, was in fact by and large an end point. The decade 614–605 BCE marks the definitive end of cuneiform culture in Assyria.

Of course we cannot know how many Assyrian scholars were amongst the survivors of this catastrophe. Very many must have died, as witnessed by the abandonment of the Nur-Šamaš home in Huzirina.[70] Some were able to start anew in Babylonia and elsewhere, trading on memorised knowledge and rescued tablets to integrate into new communities. Some men and some tablets certainly made it as far south as Uruk in the marshlands of Babylonia, where, ironically, the insurrection against Assyria had originally fomented.[71]

Independent Babylonia flourished under rebel king Nabopolassar (r. 627–605 BCE) and his son Nebuchadnezzar II (r. 606–562 BCE). As in Assyria, and in earlier times of Babylonian self-rule, there were scholars at court, attached to temples as part-time prebendary priests, and working for private clients and patrons, in many combinations.[72] Nevertheless, the sudden halving of cuneiform culture's sphere of circulation, on top of Ashurbanipal's earlier depredations, must have had a significantly deleterious effect that needed to be overcome. Quite apart from the loss of the Assyrian court and elites as a major source of patronage, there were fewer manuscripts in circulation and fewer scholarly practitioners to share (or compete) with.

Nevertheless, the sixth century BCE represents a period of prosperity and prestige for cuneiform scholarship, supported by the Babylonian dynasty's

---

[70] Seton Lloyd and Nuri Gokçe, "Sultantepe: Anglo-Turkish Joint Excavations, 1952," *Anatolian Studies* 3 (1953): 27–47.

[71] Paul-Alain Beaulieu, "The Afterlife of Assyrian Scholarship in Hellenistic Babylonia," in *Gazing on the Deep: Ancient Near Eastern and Other Studies in Honor of Tzvi Abusch*, ed. Jeffrey Stackert, Barbara Nevling Porter, and David P. Wright (Bethesda: CDL Press, 2010), 1–18; Michael Jursa, "Die Söhne Kudurrus und die Herkunft der Neubabylonischen Dynastie," *Revue d'Assyriologie* 101 (2007): 125–36.

[72] Robson, *Ancient Knowledge Networks*, chapter 4; Waerzeggers, "The Babylonian Priesthood."

patronage of temples and courtly advisors. It even survived the Cyrus the Great's conquest of Babylonia in 539 BCE. Even though the new Persian ruler was himself Zoroastrian he sought the active support of local temple communities to maintain social cohesion and economic success: essential prerequisites for a solid tax base.[73] However, Babylonian uprisings against his successors Darius I and his son Xerxes I in 521 and 484 BCE led to devastating reprisals on the urban élites who had supported the rebels.[74] Key temples in the cities of Sippar and Uruk were closed down forever, while others in Babylon, Borsippa and elsewhere were restaffed entirely with loyalists and their economic infrastructures reconfigured. Significant numbers of influential extended families disappear entirely from the cuneiform record at this point; or rather, 484 BCE is the end date of many large personal and institutional tablet collections, whether archival or learned. Tens of thousands of scholarly tablets enter the archaeological record at this point – a second survival bottleneck for the written record, as well as for the communities that produced it, just a century and a half after the first. It represents a further halving of cuneiform scholarship's sphere of circulation, at least temporarily, another dramatic loss of patronage, institutional infrastructure, and wealthy client base.

Some northern Babylonian cities, such as Sippar, seem to have lost their tradition of cuneiform literacy forever, while others, such as Babylon and Borsippa, slowly re-established temple worship and associated scholarly activity. In southern cities like Nippur, which had remained loyal to Persian rule, life apparently continued entirely unchanged. In Uruk, southern power-base of the erstwhile Babylonian dynasty, the aftermath was more complicated. Families close to the old royal line, and thus perhaps sources of new claimants to the Babylonian throne if not active supporters of known rebels, disappear from the historical

---

[73] Michael Jursa, "The Transition of Babylonia from the Neo-Babylonian Empire to Achaemenid Rule," in *Regime Change in the Ancient Near East and Egypt: from Sargon of Agade to Saddam Hussein*, ed. Harriet Crawford (Oxford: Oxford University Press, 2007), 73–94.
[74] Caroline Waerzeggers, "The Babylonian Revolts against Xerxes and the 'End of Archives,'" *Archiv für Orientforschung* 50 (2003/04): 150–73; most recently Robson, "The Socio-economics of Cuneiform Scholarship." On the increasing financial pressures on temples at this period, which may have further exacerbated the situation, see Michael Jursa, "Money-based Exchange and Redistribution: The Transformation of the Institutional Economy in First-millennium Babylonia," in *Autour de Polanyi: Vocabulaires, théories et modalités des échanges*, ed. Philippe Clancier, Francis Joannès, Pierre Rouillard, and Aline Tenu (Paris: de Boccard, 2005), 171–86; idem, "Taxation and Service Obligations in Babylonia from Nebuchadnezzar to Darius and the Evidence for Darius' Tax Reform," in *Herodot und das Persische Weltreich/Herodotus and the Persian Empire*, ed. Robert Rollinger, Brigitte Truschnegg, and Reinhold Bichler (Wiesbaden: Harrassowitz, 2011), 431–48.

record. Local elites were allowed to remain and slowly regrouped around a new temple, Reš, with a reformulated theology of the sky-god Anu, over the course of the fifth century BCE.

1.5 Survival in a Time of Scarcity: Late Babylonian Cuneiform Scholarship

This, then, is the devastating political background against which we should understand scholarly attitudes to sharing and protecting knowledge in the Uruk temple community with which this paper opens.[75] As a bastion of rebellion against Assyria in the late seventh century BCE Uruk then became both a refuge for scholars (and others?) fleeing the north and a politically powerful centre of the new Babylonian regime. For nearly a decade from 627 BCE, for instance, crown prince Nebuchadnezzar had held a sinecure as *šatammu*-bishop of the Eanna temple.[76] After the Achaemenid conquest, as discontent with the new imperial realities grew, Eanna was a focal point for increasing tensions between crown and cult. The temple was wound down early in Darius's reign, leading perhaps to some prominent elements of Uruk society supporting further rebellions on the accession of his son Xerxes. The young king certainly identified Uruk as a source of trouble, removing key families from city power – and perhaps removing them altogether – in 484 BCE.

The Ekur-zakir, Hunzu, Šangu-Ninurta and Sin-leqe-unninni families all survived these tumultuous decades in Uruk. When Alexander the Great marched into Babylon in 330 BCE, and then when his former general Seleucus eventually consolidated his own rule two decades later, it is no wonder that Uruk chose discretion over valour, obscurity over proximity to power. From late Achaemenid and into Seleucid times the city's *āšipu*s and *kalû*s knew very well that they could expect no royal patronage, for themselves or their temple. They were dependent entirely on income from part-time prebendary rights and from personal consultations for

---

[75] Stevens, "Secrets in the Library"; Robson, "The Production and Dissemination"; eadem, "Tracing Networks of Cuneiform Scholarship"; "The Socio-economics of Cuneiform Scholarship"; Robson and Stevens, "Scholarly Tablet Collections in First-Millennium Assyria and Babylonia." For more detail on the differing fates of Babylonian scholarly communities, especially those of Borsippa and Uruk, post-484 BCE see Robson, "The Socio-economics of Cuneiform Scholarship," on which the following paragraphs are based, with extensive references to further literature.

[76] Jursa, "Die Söhne Kudurrus," 131.

healing, horoscopes, and the like.[77] But they could count on custom and respect from only a proportion of the urban community: Jews, Zoroastrians, Greeks, and other cultures were all part of city life now. It is impossible to tell whether or not Babylonian traditionalists still made up the majority of Uruk's population in the fourth and third centuries BCE but they certainly did not constitute the near-monopoly of earlier times.

In this light, then, the Uruk scholars' motivations for operating a "distributed library" of shared and protected knowledge must have been rather different to those of their Assyrian precursors, even if their strategies appear similar. As in the seventh-century urban scholarly communities examined above, the Uruk men acquired tablets from as far afield as Nippur, Kutha, and Der, as well as from others in their immediate communities.[78] They borrowed and returned after "hasty excerpting," they worried about the risks of loaning works out themselves, and summoned their personal gods to protect them. In particular, we can understand better the particular protective measures that Stevens describes for compositions closest to individual scholars' livelihoods.[79] Recall from the introduction the astronomical calculations drawn up in April 191 BCE, whose colophon utilises no less than four different protective strategies:

> Tablet of Anu-belšunu, *kalû* of the god Anu, son of Nidintu-Anu, descendant of Sin-leqe-unninni, Urukean. Hand of Anu-[aba-uter, his son, *ṭupšar Enūma*] *Anu Ellil*, Urukean.
>
> Uruk, Nisannu (month I), year 1 21, Antiochus [was king].
>
> He who reveres the gods Anu, Ellil and Ea [shall not take] it away by theft(?). Ephemeris, wisdom of Anu-ship, secret of the [great] gods, treasure of the scholars. The learned may show [the learned]; the unlearned may not [see. Taboo] of Anu, Ellil [and Ea, the great gods].[80]

The *āšipu*s and *kalû*s of Late Babylonian Uruk were responding to several types of threats through scarcity. The first was scarcity of royal patronage. That meant they could comfortably discount the possibility of large-scale confiscation of tablets à la Ashurbanipal, but the community memory of Xerxes's destruction of scholarly families, communities and temples must have remained raw. Less drastically but

---

77 Robson, "The Socio-economics of Cuneiform Scholarship," 466.
78 Robson, "Tracing Networks of Cuneiform Scholarship," 157.
79 Stevens, "Secrets in the Library."
80 pa-lih 21 50 u 40 ina šur$^{?}$ qa$^{?}$ [la TÙM]-šú | a-ru-ú né-me-qí ^{d}60-ú-tú ⌈AD.HAL DINGIR⌉.[MEŠ GAL.MEŠ] | MÍ.ÙRI ^{lú}um-man-nu ^{lú}ZU$^{-ú}$ ana [^{lú}ZU$^{-ú}$] | li-kal-lim la ^{lú}ZU$^{-ú}$ nu [im-mar ik-kib] | ^{d}a-⌈nù⌉$^{-d}$EN.LÍL ⌈ù⌉ [^{d}é-a DINGIR.MEŠ GAL.MEŠ] (Neugebauer, *Astronomical Cuneiform Texts*, no. 135U rev. 12–16; cf. Stevens, "Secrets in the Library," 252 no. 45).

more immediately, the diversification of personal beliefs and religious practices meant that temple worshippers and clients for divination and healing were in ever shorter supply, while inheritance customs encouraged prebendary shares in temple income to be split into smaller and smaller parts. Meanwhile, the Assyrian and Achaemenid "survival bottlenecks" had taken untold numbers of scholarly works out of circulation, meaning that the textual basis of their professions was ever harder to come by.

Cynthia Jean has tracked the availability of compositions listed in the classic *Āšipu's Handbook*.[81] In seventh-century Assur, where junior *āšipu* Kiṣir-Aššur of the Baba-šumu-ibni family made a copy of it, his family owned about half of the hundred or so compositions listed there (and maybe more if we take long-vanished writing-boards and unexcavated areas of the house into account).[82] In late fifth-century Uruk junior *āšipu* Anu-ikṣur of the Šangu-Ninurta family also made a copy, but his family had about half that number again.[83] It must have been painfully obvious to him how many of the key works of his profession were no longer in circulation. In these straitened circumstances it was more important than ever before to hoard what one had, and to share only with a trusted few. Right until the last generation of scholarly activity in Uruk, in the mid-second century BCE, copyists were still writing on their tablets, "He who reveres the god Anu shall not carry it off."[84]

Meanwhile, the story of cuneiform scholarship in the city of Babylon in the centuries after the anti-Achaemenid revolts is still to be pieced together. However, we do know that Xerxes saw Marduk's temple Esangila as the epicentre of the Babylonian independence movement, and that his reprisals included the decommissioning of its ziggurat, the dismantling of its prebendary system, and wholesale replacement of its senior personnel. There was some rapprochement with political power under Alexander the Great and the early Seleucids, when, for

---

81 Cynthia Jean, *La magie néo-assyrienne en contexte: Recherches sur le métier d'exorciste et le concept d'*āšipūtu (Helsinki: The Neo-Assyrian Text Corpus Project, 2006), 165–67.
82 KAR 44 (ed. Geller, "Incipits and Rubrics").
83 Ernst von Weiher, *Spätbabylonische Texte aus Uruk, 5te Band*, Ausgrabungen der Deutschen Forschungsgemeinschaft in Uruk-Warka, Endberichte 13 (Mainz: von Zabern, 1998), no. 321; Philippe Clancier, "Le manuel de l'exorciste d'Uruk," in *Et il y eut un esprit dans l'Homme: Jean Bottéro et la Mésopotamie*, ed. Xavier Faivre, Brigitte Lion, and Cécile Michel (Paris: De Boccard, 2009), 105–17.
84 pa-lih ᵈ60 là ⌈TÙM⌉-šú (Jan J. A. van Dijk and Werner R. Mayer, *Texte aus dem Rēš-Heiligtum in Uruk-Warka* [Berlin: Mann, 1980], no. 89 rev. 9), a list of historical kings and their scholarly advisor drawn up by Anu-belšunu's eponymous grandson in 165 BCE (Alan Lenzi, "The Uruk List of Kings and Sages and Late Mesopotamian Scholarship," *Journal of Ancient Near Eastern Religions* 8 [2008]: 137–69).

instance, Esangila's *šatammu*-bishop Berossos (Babylonian Bel-re'ušu?) supposedly dedicated his famous Greek-language history Babyloniaca to Antiochus I in about 280 BCE.[85] But this moment of cultural exchange is perhaps indicative of a larger sense that cuneiform scholarship was no longer viable as an independent body of knowledge, and needed to be shared more widely, in new languages.[86] It is probably also in the third century BCE that some elements of Babylonian temple astronomy started to filter into the Greek tradition.[87]

The very last known cuneiform scholarly community functioned in Babylon over the period c. 150–50 BCE. Seleucid power and territory had been waning since the early second century BCE, under pressures from Rome to the west, Ptolemaic Egypt to the south, and the Parthians to the east. In 141 BCE, the royal city of Seleuceia-on-Tigris fell to the Parthians, who then set up a new imperial centre just 5 km away. Like the Seleucids before them, the new rulers of Babylonia rejected Babylon, some 65 km to the south, as a royal residence but allowed the city and its now much diminished temple, Esangila, to continue in existence.

Although we are lacking stratigraphically excavated, published archives from this period, informally recovered tablets from Victorian expeditions show that even at this late date Esangila was still a locus of scholarly as well as cultic activity. The famous Astronomical Diaries, the latest of which dates to 61 BCE,[88] were produced under its auspices, systematically recording a wealth of celestial and meteorological observational data, as well as increasingly frequent and extensive records of military, religious and political events.[89] The Diaries regularly mention sacrifices and rituals in Esangila until at least 78 BCE,[90] while a small group of letters and legal documents of the temple scholars who made the

---

85 Johannes Haubold, Giovanni Lanfranchi, Robert Rollinger, and John Steele, *The World of Berossos* (Wiesbaden: Harrassowitz, 2013).
86 John Dillery, *Clio's Other Sons: Berossus and Manetho* (Ann Arbor: University of Michigan Press, 2015).
87 Alexander Jones, "Transmission of Babylonian Astronomy to Other Cultures," in *Handbook of Archaeoastronomy and Ethnoastronomy*, ed. Clive N. Ruggles (New York: Springer, 2015), 1877–81.
88 Abraham J. Sachs and Hermann Hunger, *Astronomical Diaries and Related Texts from Babylonia, Volume III: Diaries from 164 B.C. to 61 B.C.* (Vienna: Austrian Academy of Sciences, 1996), no. 62.
89 Reinhart Pirngruber, "The Historical Sections of the Astronomical Diaries in Context: Developments in a Late Babylonian Scientific Text Corpus," *Iraq* 75 (2013): 197–210.
90 For instance, in May 78 BCE, "(the) governor of Babylon entered Babylon. That day, the *šatammu*-bishop of Esangila and the Babylonians, the *kiništu*-assembly of Esangila, provided [1 bull] and 2 (sheep) sacrifices at the Gate of the Prince's Son in Esangila as an offering for this governor of Babylon" (Sachs and Hunger, *Astronomical Diaries*, no. 77A obv. 26'–27').

Diary observations dates to 127–103 BCE.[91] Most prominent amongst them was perhaps Itti-Marduk-balaṭu, son of Iddin-Bel, "gardener, city supervisor(?), overseer of the gods' temples, ṭupšar Enūma Anu Ellil, who had previously attended(?) Hyspaosines the king."[92] On 30 May, 127 BCE, the šatammu-bishop and temple assembly formally agreed that his two sons should take over his observational and calculational work. The latest surviving administrative records from Esangila were drawn up for a man called Rahim-Esu in 94–93 BCE, who essentially served as one of the temple's bankers: he managed its income and paid its salaries and expenses, running this operation as a profit-making business rather than as a direct employee.[93]

Exactly contemporary with these records are scholarly texts written by members of three families associated with the temple, who also interrelated with each other: namely the descendants of Egibatila, Mušezib, and Nanna-utu. Babylonian scholarly lineages, along with Esangila's prebendary system of priestly duties and privileges, had largely been wiped out by Xerxes in 484 BCE[94]; these families are amongst the few who survived or emerged in the aftermath. The Mušezib family had been central to the development of mathematical astronomy in the late fourth century BCE and continued to be members of the observational community in the late second, as witnessed by the letters and legal documents

---

91 Gilbert J. P. McEwan, *Priest and Temple in Hellenistic Babylonia* (Wiesbaden: Steiner, 1981), 17–21; Robartus J. van der Spek, "The Babylonian Temple during the Macedonian and Parthian Domination," *Bibliotheca Orientalis* 42 (1985): 541–62, at 548–55; Michael Jursa, *Neo-Babylonian Legal and Administrative Documents: Typology, Contents and Archives* (Münster: Ugarit, 2005), 75; Johannes Hackl, "Materialien zur Urkundenlehre und Archivkunde der spätzeitlichen Texte aus Nordbabylonien" (PhD diss., Vienna University, 2013), 461–71.

92 ^{m}KI-^{d}ŠÚ-DIN ^{lú}GAL.DÙ | ‹šá› UGU IRI ^{lú}up-pu-de-tú šá É.MEŠ DINGIR.MEŠ | ‹^{lú}›UMBISAG U$_4$ AN.NA ^{d}EN.LÍL.LÁ A «LÚ» šá ^{m}MU-^{d}EN | ša i-na IGI-ma a-na Á as-pa-si-né-e LUGAL | DÙ (Theophilus G. Pinches, "A Babylonian Tablet Dated in the Reign of Aspasine," *Babylonian and Oriental Record* 4 [1896]: 131–35 obv. 9–13, cf. van der Spek, "The Babylonian Temple," 549–551).

93 Gilbert J. P. McEwan, "Arsacid Temple Records," *Iraq* 43 (1981): 131–43; Robartus J. van der Spek, "Cuneiform Documents on Parthian History: The Rahimesu Archive – Materials for the Standard of Living," in *Das Partherreich und seine Zeugnisse/The Arsacid Empire: Sources and Documentation*, ed. Josef Wiesehöfer (Stuttgart: Steiner, 1998), 205–58; Jursa, *Neo-Babylonian Legal and Administrative Documents*, 75–76; Johannes Hackl, "New Additions to the Rahimesu Archive: Parthian Texts from the British Museum and the World Museum Liverpool," in *Silver, Money and Credit: A Tribute to Robartus J. van der Spek on the Occasion of his 65th Birthday*, ed. Kristin Kleber and Reinhard Pirngruber (Leiden: Nederlands Instituut voor het Nabije Oosten, 2016), 87–106.

94 Hackl, "Materialien zur Urkundenlehre," 393.

mentioned above.⁹⁵ Some of their astronomical work survives, as well as a copy of Tablet X of The Epic of Gilgamesh, written for Itti-Marduk-balaṭu by one of his sons, Bel-ahhe-uṣur.⁹⁶ A few members of the Egibatila and Nanna-utu families also learned calculational astronomy, one of them studying with Bel-ahhe-uṣur's relative Marduk-šapik-zeri Mušezib.⁹⁷ Meanwhile Nabu-balassu-iqbi, descendant of Egibatila, specialised in commentaries on various types of omen compilations.⁹⁸ Three generations of *kalû*-lamenters from the Nanna-utu family, by contrast, wrote out long ritual laments "excerpted for singing" in Emesal, the ancient liturgical dialect of Sumerian, with interlinear Akkadian translations.⁹⁹ One of them also trained a member of the Egibatila family in lamentation, which suggests that some of them too were *kalûs*.¹⁰⁰

Nearly ninety scholarly tablets have so far been assigned to men of these three families, nearly half of which have (partially) surviving colophons dating to between 137 and 49 BCE. Apart from the fourteen tablets *ana zamāri nashi*

---

95 Joachim Oelsner, "Von Iqīšâ und einigen anderen spätgeborenen Babyloniern," in *Studi su vicino Oriente antico dedicati alla memoria di Luigi Cagni*, ed. Simonetta Graziani (Napoli: Istituto Universitario Orientale, 2000), 797–813, at 802–10; Eleanor Robson, *Mathematics in Ancient Iraq: A Social History* (Princeton: Princeton University Press, 2008), 221–26; Mathieu Ossendrijver, *Babylonian Mathematical Astronomy: Procedure Texts* (New York: Springer, 2012), 8 n. 44.
96 Parthian-period scholarly tablets with Mušezib colophons include Andrew R. George, *The Babylonian Gilgameš Epic: Introduction, Critical Edition and Cuneiform Texts* (Oxford: Oxford University Press, 2003), 114 source b (Gilgamesh); Grant Frame and Andrew R. George, "The Royal Libraries of Nineveh: New Evidence for King Ashurbanipal's Tablet Collecting," *Iraq* 67 (2004): 265–84, at 268 (literary letter); and Neugebauer, *Astronomical Cuneiform Texts*, no. 123Zk (astronomy).
97 Neugebauer, *Astronomical Cuneiform Texts*, nos. 18Zq, 122Zo, 420+821Zld, 611+822Zm.
98 The Cuneiform Commentaries Project, directed by Eckart Frahm at Yale University, gives a full catalogue, bibliography and online edition of Nabu-balassu-iqbi's commentaries (http://ccp.yale.edu/catalogue?ccp=&scribe=Nabu-balassu-iqbi, accessed 1 September 2016).
99 George Reisner, *Sumerische-Babylonische Hymnen nach Thontafeln Griechischer Zeit* (Berlin: Spemann, 1896), nos. 3, 5, 10, 15, 18, 19, 20a, 25, 27, 28, 36, 44, 45, 46, 49, 51, 53, and 55; Ira Spar and Wilfred G. Lambert, eds., *Literary and Scholastic Texts of the First Millennium BC*, Cuneiform Texts in the Metropolitan Museum of Art 2 (New York: The Metropolitan Museum of Art, 2005), nos. 2, 8, and 15. Preliminary online edition by the Bilinguals in Late Mesopotamian Scholarship project directed by Steve Tinney at the University of Pennsylvania (http://oracc.org/blms, accessed 1 September 2016).
100 As I argue elsewhere, *kalûs* were often secondarily *ṭupšar Enūma Anu Ellil*, not only in Hellenistic Uruk where they are particularly well attested, but throughout the first millennium BCE in both Assyria and Babylonia; Eleanor Robson, "Who Wrote the Babylonian Astronomical Diaries?" in *Keeping Watch in Babylon: from Evidence to Text in the Astronomical Diaries*, ed. Johannes Haubold, John Steele, and Kathryn Stevens (Boston: Brill, forthcoming).

"excerpted for singing,"[101] many more state explicitly that they have been copied from other sources – from one Belšunu's house, from the nearby city of Borsippa, even from *magallatu* (leather rolls) from Babylon.[102] Ten have (partially) surviving protective formulae, across all four genres: astronomy, commentary, literature, and liturgy. Nabu-mušetiq-uddi Mušezib warns, "He who reveres the god Šamaš must not erase my handiwork."[103] Nabu-balassu-iqbi Egibatila invokes the god Nabu, fully and inventively:

> [He who reveres] the god Nabu should greatly, greatly guard and treasure (this tablet); [he may] not [show] it to anyone who is not the son of a work-master.[104]

It appears that the concept of the "learned" and the "unlearned" was now obsolete; it has been replaced with a social signifier. Although the exact meaning of the term *mār bēl dulli*, literally "son of a work-master," is unclear at this late period, it is perhaps related to the earlier *mār banê*, widely used into Hellenistic times. Liter-

---

101 E.g., ⌜ana⌝ DU$_{12}$ ⌜ZI⌝^{hi} | IM.GÍD.DA ^{md}EN-A-⌜MU A⌝ šá ^{mdr}IDIM⌝-DIN^{su}-E ⌜A⌝ | ^{md}nanna-u$_3$-tu ŠU ^{md}EN-MU-NA A ⌜šá⌝ | ^{m}KI-^{d}ŠÚ-TIN A ^{m}e$_4$-gi$_7$-ba-tìl-la ⌜TIN.TIR⌝[^{ki}] "Extracted for singing. Exercise tablet of Bel-apal-iddin, son of Ea-balassu-iqbi, descendant of Nanna-utu. Handiwork of Bel-šumu-lišir, son of Itti-Marduk-balaṭu, descendant of Egibatila, Babylon" (Reisner, *Sumerische-Babylonische Hymnen*, no. 3 rev. 10'–13', a bilingual *ballangu*-liturgy).
102 SUMUN-šú ina É ^{m}EN-šu-nu [...]-⌜x⌝ | ^{im}DUB ^{m}U-A-[MU A šá ^{md}IDIM]-DIN-^{su}-E | A šá ^{mdr}nanna⌝-[ù-tu ...] "Its original is from the house of Belšunu [...]. Tablet of Bel-apla-[iddin, son of Ea]-balassu-iqbi, descendant of Nanna-[utu ...]" (Spar and Lambert, *Literary and Scholastic Texts*, no. 15, a bilingual *šuillakku*-prayer to the god Ninurta); ⌜LIBIR.RA-šú⌝ TA muh-hi IM.GÍD.DA SUMUN GABA.RI bar-sìp^{ki} SAR-ma IGI.TAB | IM.GÍD.DA ^{md}NÀ-DIN^{su}-E A šá ^{md}AMAR.UTU-NUMUN-DÙ A ^{md}egi-ba-ti-la | ŠU.MIN ^{md}NÀ-MU-SI.SÁ DUMU-šú "Its original is from an old exercise tablet of Borsippa, copied and checked. Exercise tablet of Nabu-balassu-iqbi, son of Marduk-zer-ibni, descendant of Egibatila. Handiwork of Nabu-šum-lišir, his son" (Cyril. J. Gadd, *Cuneiform Texts from Babylonian Tablets, &c., in the British Museum, Part XLI* [London: The British Museum, 1931] pl. 31 rev. 36–38; cf. pl. 32 rev. 24–26, both commentaries on the omen series Šumma Ālu); DUB šá EGIR-šú ... | ... ina ^{kuš}ma-gal-lat GABA.RI E^{ki} [SAR ^{im}DUB] | ⌜^{rm}^{d}NÀ-DIN^{su}-E A šá ^{md}AMAR.UTU-NUMUN-DÙ A ^{md}e$_4$-[gi$_7$-ba-ti-la] "Tablet whose continuation (quotes the first line) is [written] on a leather roll, a manuscript from Babylon. [Tablet of] Nabu-balassu-iqbi, son of Marduk-zer-ibni, descendant of Egibatila" (Ernst Weidner, "Ein Kommentar zu den Schlangen-Omina," *Archiv für Orientforschung* 21 [1966]: 46, pl. 10 rev. 38–40, cf. http://ccp.yale.edu/P461205 rev. 5'–8', both commentaries on Šumma Ālu; http://ccp.yale.edu/P433502 rev. 1'–4', commentary on sacrificial omens, accessed 1 September 2016).
103 pa-lih 20 ŠU.MIN là í-paš$_x$(GÍN)-⌜šiṭ⌝ (Frame and George, "The Royal Libraries of Nineveh," 368 rev. 23, a literary letter, on which see further below with note 108).
104 [GIM LIBIR-šú ^{md}U$_4$.U$_4$.U$_4$].⌜U$_4$⌝.U$_4$.U$_4$.U$_4$.U$_4$.U$_4$-DIN^{su}-E A šá ^{md}AMAR.UTU-NUMUN-DÙ <A> ^{md}e$_4$-gi$_7$-ba-ti-la | ⌜SAR⌝-ma ib-ri | [pa-liḫ ^{d}U$_4$.U$_4$.U$_4$].⌜U$_4$⌝.U$_4$.U$_4$.U$_4$.U$_4$.U$_4$ ma-diš ma-diš li$_6$-ṣur li$_6$-šá-qir al-la DUMU EN du-ul-la ⌜là⌝ [ú-kal-lam] (Spar and Lambert, *Literary and Scholastic Texts*, no. 69 rev. 3'–5', commentary on a medical text).

ally translated as "son of the good," this phrase is explained by Michael Jursa as a "non-serf head of a household (loosely) affiliated to the temple."¹⁰⁵ The word *dullu*, "work," was a common term for the (now obsolete) labour-taxation paid by temple communities until the early Achaemenid period.¹⁰⁶ If this identification of *mār bēl dulli* is correct, then the permissible sphere of circulation for scholarly writings has shifted from the highly cuneiform-literate to the temple community: a tacit acknowledgement that cuneiform was no longer meaningful in the world beyond?¹⁰⁷

Whatever this phrase might signify, unlike their Assyrian and Late Babylonian forebears, neither man articulates what the consequent divine punishment might be. Moreover most scholars in their circle write an even more perfunctory abbreviation of this standard phrase, omitting the offending action itself. "He who reveres the gods Šamaš and Marduk," declare the Egibatila and Nanna-utu men; "He who reveres the gods Bel and Beltiya," invoke the Mušezibs.¹⁰⁸ Just *what*, exactly, the reverent man is supposed to do with the tablet – return, protect, treasure, not remove, not lose – and under what penalty, is never declared.

These phrases, as well as the explicitly stated copying habits described above, make it clear that the scholars of Parthian Babylon expected others to have access to their writings. Yet the lacklustre nature of their protective formulae suggests that they did not anticipate much inappropriate human interest in their writings, and/or did not really count on the gods to provide appropriate protection. Indeed, they seem to have given up on the protective habit entirely by the first century BCE.¹⁰⁹ Certainly, at this very late juncture in Babylonian culture, there must have

---

105 Michael Jursa, "Labor in Babylonia in the First Millennium BC," in *Labor in the Ancient World*, ed. Piotr Steinkeller and Michael Hudson (Dresden: ISLET, 2015), 345–96, at 351.
106 Jursa, "Taxation and Service Obligations," 442; idem, "Labor in Babylonia," 352.
107 Independently Johannes Hackl, "Language Death and Dying Reconsidered: The Role of Late Babylonian as a Vernacular Language," Imperium and Officium Working Papers, 2011. Vienna: http://iowp.univie.ac.at/, 16 posits the second century BCE – exactly the period we are discussing here – as the point at which Akkadian probably died out as a vernacular language in favour of Aramaic.
108 E.g., ⌈pa⌉-lih ᵈUTU u ᵈAMAR.UTU (http://ccp.yale.edu/P461205 rev 9', see note above); pa-lih ᵈEN u ᵈGAŠAN-ia (George, *The Babylonian Gilgameš Epic*, 114 source b rev. ii 18', see above with note 96). The following, damaged sign that George, *The Babylonian Gilgameš Epic*, 114 reads as GUR, the logogram for *târu* "to return," is to my mind more likely to be ⌈E⌉[ᵏⁱ] "Babylon" (cf. e.g., Spar and Lambert, *Literary and Scholastic Texts*, no. 2 rev. 20').
109 The latest dated tablet known to me that bears the phrase "He who reveres the gods Šamaš and Marduk" is DT 35, a commentary on the ominous calendar *Iqqur Īpuš*, written by a member of the Egibatila family in 103 BCE (http://ccp.yale.edu/P461300). At least eight tablets written by scholars in the Egibatila circle post-date it, the latest being a calculated table of full moons from 49 BCE by a member of the Nanna-utu family (Neugebauer, *Astronomical Cuneiform Texts*, no. 18Zq).

been very few other cuneiform literate communities around. But if they were merely going through the motions, for custom's sake, why bother at all?

Fascinatingly, this group of scholars were still acutely aware of Ashurbanipal's long-ago plundering of the scholarship of northern Babylonia, as witnessed by two literary letters copied by members of the Mušezib and Egibatila families.[110] One letter purports to be from "the obedient citizens of Borsippa," promising to obey the king's command to "Write out all the scribal learning in the property of the god Nabu and send it to me!" and referring him to Esangila for one particular text – a Sumerian vocabulary – that is not in their possession.[111] The other letter is a longer response to a similar royal request for "all the scribal [learning, as much as there is, that is in the possession] of the great lord Marduk, my lord." In this composition, twelve named scholars from Babylon offer to write down all that is "stored in their minds like goods piled in a magazine" in exchange for silver and political favour.[112]

The historicity of the original letters is still hotly debated, but what matters here is that in Babylon, over half a millennium later, the group memory of this event was still current. However, in this late recounting, no original tablets left Babylonia for Nineveh (though we have seen in the previous section that this was not the case) and no scholars were chained up in the royal palace and forced to work. Instead they offered to transfer their knowledge from memory onto writing boards in return for royal respect and reward. This rose-tinted retelling was a

---

110 Frame and George, "The Royal Libraries of Nineveh"; cf. Eckart Frahm, "On Some Recently Published Late Babylonian Copies of Royal Letters," *Nouvelles Assyriologiques Brèves et Utilitaires* 43 (2005): 43–46.

111 bar-sìp^{ki}-MEŠ sa-an-⌈qu-tú⌉ a-na LUGAL EN-šu-nu ú-ta-ru-⌈šú⌉ na-áš-par-tu₄ šá ⌈iš-ṭu-ru⌉ | um-ma kul-lat ^{lú}DUB.SAR^{tú} ⌈šá ŠÀ⌉ NÍG.GA ^{d}NÀ EN-ía šu-ṭu-ra-a' šu-bil-la-ni | šul-li-i'-a [na-áš]-par-⌈tu₄⌉ "The obedient citizens of Borsippa will return (i.e., fulfill) to their king the commission that he wrote, as follows: 'Write out all the scribal learning in the property of the god Nabu, my lord, and send it to me! Fulfill the commission!'" (Frame and George, "The Royal Libraries of Nineveh," 268 obv. 8–10). The colophon reads: GIM <SUMUN>-šú SAR-ma IGI.TAB u IGI.KÁR ^{im}DUB ^{md}EN-TIN^{su} A šá ^{md}NÀ-DIB-U₄.DA A ^{m}mu-še-zib | ŠU.MIN ^{md}NÀ-DIB-U₄.DA A-šú "Copied and checked according to its <original>. Tablet of Bel-uballissu, son of Nabu-mušetiq-uddi, descendant of Mušezib. Handiwork of Nabu-mušetiq-uddi, his son" (rev. 22–23, and see note 103 above for its continuation).

112 kul-lat ^{lú}DUB.[SAR^{tú} ... šá] | [ŠÀ] ⌈NÍG⌉.GA ^{d}AMAR.UTU ^{d}EN GAL^{ú} EN-iá (Frame and George 2004:273 obv 9–10); 12 ^{lú}UM.ME.A.MEŠ an-nu-tú ... | ... [kul-lat ^{lú}DUB.SAR^{tú}] | [šá] i-hi-ṭu-ú ib-ru-ú GIM gu-ru-⌈un⌉-né-e a-na kar-ši-šú-nu kam-su "these 12 scholars ... [all of scribal learning] that they have read and checked, stored in their minds like goods piled in a magazine" (obv. 13–14). The remains of the colophon are restored by Frahm, "On Some Recently Published Late Babylonian Copies," 45) to read: [IM.GÍD.DA ^{md}EN-MU-SI.SÁ DUMU šá ^{m}KI-^{d}AMAR.UTU]-DIN DUMU ^{md}e₄-gi₇-ba-⌈ti⌉-[la ...] "[Exercise tablet of Bel-šum-lišir, son of Itti-Marduk]-balaṭu, descendant of Egibatila [...]".

reimagining of a time in which cuneiform scholarship was still in high demand, when even the world's most powerful king treated the learned with the deference they felt they deserved but had lost long ago.

1.6 Conclusions

The costs and benefits to sharing or concealing written knowledge in cuneiform culture were weighted differently in different times and places, according to the opportunities and pressures of the moment. Most simply put, the higher up the social scale the less need there was for scholarly reciprocity. Members of the Nur-Šamaš and Baba-šumu-ibni family circles in seventh-century Huzirina and Assur took care to acknowledge their sources and copyists (who mostly had junior status), and to return tablets borrowed from others. In turn they expected the same courtesies from others in their intellectual communities. Without such a formally encoded etiquette for sharing and protecting, any one individual's chances of access to the written word were substantially diminished. By contrast their courtier contemporaries, the descendants of Gabbu-ilani-ereš and their colleagues, did not credit their scribes and did not expect tablets to be borrowed or copied by others. Stored in the inner courtyard of Ezida on the royal citadel, under the watchful eye of Nabu himself, their writings were as safe as could possibly be. Only the king himself could assert any claim on them. And this was part of a much larger, longer-term royal attempt to centralise and monopolise scholarly knowledge. Focusing overwhelmingly on divination, ritual, healing, and prayer, Ashurbanipal's vast personal tablet collection aimed not only to diminish other humans' access to learning but to maximise his own ability to predict and control the gods' will.

However, even if – or rather, precisely because – in reality sharing and protecting of written knowledge was socially asymmetrical, it was not possible to admit that truth in practice. Hence the euphemistic worries expressed about "the unlearned" gaining inappropriate access to writings which, as we saw at the very start of this article, would be have been utterly incomprehensible to all but a handful of the highly cuneiform literate. We have also seen how tightly individual families held on to scholarly roles across the generations, whether as royal advisors like the Gabbu-ilani-ereš men, or temple ritualists like the Sin-leqe-un-ninnis. There was no real threat of untrained outsiders accessing sufficient professional instruction, never mind sufficient social status, to set themselves up as rival *āšipu*s or *kalû*s to the long-established urban dynasties.

Rather, as we have seen, scholarly communities were most at risk from state-level threats because cuneiform scholarship was seen to be powerful, and therefore highly politicised and threatening. For Ashurbanipal in the mid-seventh century BCE, the means to read and understand the gods' will should be the king's above all, and even if he did not intend to deprive others entirely of those means, he insisted on unprecedentedly complete access to the writings that enabled communication with the divine. War against his brother Šamaš-šumu-ukin in Babylonia gave him the perfect opportunity to pursue that plan but also, ultimately, led to its and the empire's collapse, gravely imperilling the survival of cuneiform scholarship in the process. Conversely, 150 years later, Darius and Xerxes were not believers in the Babylonian gods but saw the temples as a source of taxation revenue on the one hand and of political rebellion on the other. Shutting down the latter while maintaining the former entailed the removal of local centres of resistance, both institutional and familial. The scholars and temples of northern Babylonia were again grievously affected. Over the course of a century and a half, cuneiform scholarship's sphere of circulation had halved and halved again.

The scholarly community, ever resilient, rebuilt and reconfigured itself once more. But henceforth it would be wary of too much engagement with royal power, which could veer from the over-invested to the violently hostile. From the fifth century onwards, in the absence of kingly patronage, cuneiform scholarship's real struggle was to find local validation and income, whether through temple affiliation or private clientele. But urban populations now had more choice of divine authority than ever before, and traditional Babylonian learning had to compete with new ways of thinking from both east and west. Worries about protecting and sharing written knowledge were perhaps most acute in the late Achaemenid and early Seleucid periods. But eventually, over the course of the third and second centuries BCE, the shrinking community of the cuneiform-literate accepted that they had lost the battle for status and influence amongst their fellow city dwellers. One strategy was to share their learning more widely, in alphabetic scripts, via mechanisms and to readerships that we still do not fully understand. But on the street and in the (emptying) temple there was now little interest in what these erstwhile experts did and thought, compared to the glory days of cuneiform culture, few fellow-travellers with whom to share it and therefore very little need to protect their traditional writings in the once customary way.

References

Paul-Alain Beaulieu, "The Afterlife of Assyrian Scholarship in Hellenistic Babylonia," in *Gazing on the Deep: Ancient Near Eastern and Other Studies in Honor of Tzvi Abusch*, ed. Jeffrey Stackert, Barbara Nevling Porter, and David P. Wright (Bethesda: CDL Press, 2010), 1–18.

Carl Bezold, *Catalogue of the Cuneiform Tablets in the Kouyunjik Collection of the British Museum, Volume 5* (London: The British Museum, 1899).

Rykle Borger, "Bemerkungen zu den akkadischen Kolophonen," *Welt des Orients* 5 (1969–70): 165–71.

Philippe Clancier, "Le manuel de l'exorciste d'Uruk," in *Et il y eut un esprit dans l'Homme: Jean Bottéro et la Mésopotamie*, ed. Xavier Faivre, Brigitte Lion, and Cécile Michel (Paris: De Boccard, 2009), 105–17.

Stephen W. Cole and Piotr Machinist, *Letters from Assyrian and Babylonian Priests to Kings Esarhaddon and Assurbanipal*, State Archives of Assyria 13 (Helsinki: Helsinki University Press, 1998).

Manfred Dietrich, *The Neo-Babylonian Correspondence of Sargon and Sennacherib*, State Archives of Assyria 17 (Helsinki: Helsinki University Press, 2003).

Jan J. A. van Dijk and Werner R. Mayer, *Texte aus dem Rēš-Heiligtum in Uruk-Warka* (Berlin: Mann, 1980).

John Dillery, *Clio's Other Sons: Berossus and Manetho* (Ann Arbor: University of Michigan Press, 2015).

F. Mario Fales and J. Nicholas Postgate, *Imperial Administrative Records, Part I: Palace and Temple Administration*, State Archives of Assyria 7 (Helsinki: Helsinki University Press, 1992).

F. Mario Fales and J. Nicholas Postgate, *Imperial Administrative Records, Part II: Provincial and Military Administration*, State Archives of Assyria 11 (Helsinki: Helsinki University Press, 1995).

Jeanette C. Fincke, "The Babylonian Texts of Nineveh: Report on the British Museum's Ashurbanipal Library Project," *Archiv für Orientforschung* 50 (2003/04): 111–49.

Eckart Frahm, "Nabû-zuqup-kenu, das Gilgameš-Epos und der Tod Sargons II," *Journal of Cuneiform Studies* 51 (1999): 73–90.

Eckart Frahm, "On Some Recently Published Late Babylonian Copies of Royal Letters," *Nouvelles Assyriologiques Brèves et Utilitaires* 43 (2005): 43–46.

Eckart Frahm, "Keeping Company with Men of Learning: The King as Scholar," in *The Oxford Handbook of Cuneiform Culture*, ed. Karen Radner and Eleanor Robson (Oxford: Oxford University Press, 2011), 508–33.

Grant Frame, *Rulers of Babylonia: From the Second Dynasty of Isin to the End of Assyrian Domination (1157–612 BC)*, Royal Inscriptions of Mesopotamia: Babylonian Periods 2 (Toronto: University of Toronto Press, 1995).

Grant Frame and Andrew R. George, "The Royal Libraries of Nineveh: New Evidence for King Ashurbanipal's Tablet Collecting," *Iraq* 67 (2004): 265–84.

Cyril. J. Gadd, *Cuneiform Texts from Babylonian Tablets, &c., in the British Museum, Part XLI* (London: The British Museum, 1931).

Markham J. Geller, "Fragments of Magic, Medicine and Mythology from Nimrud," *Bulletin of the School of Oriental and African Studies* 63 (2000): 331–39.

Markham J. Geller, "Incipits and Rubrics," in *Wisdom, Gods and Literature: Studies in Assyriology in Honour of W.G. Lambert*, ed. Andrew R. George and Irving L. Finkel (Winona Lake: Eisenbrauns, 2000), 225–58.

Markham J. Geller, "Look to the Stars: Babylonian Medicine, Magic, Astrology and Melothesia," Max Planck Institute for the History of Science Preprints 401 (Berlin: Max Planck Institute for the History of Science, 2010).

Andrew R. George, *The Babylonian Gilgameš Epic: Introduction, Critical Edition and Cuneiform Texts* (Oxford: Oxford University Press, 2003).

Petra D. Gesche, *Schulunterricht in Babylonien im ersten Jahrtausend v. Chr.*, Alter Orient und Altes Testament 275 (Münster: Ugarit, 2000).

Oliver R. Gurney and Jacob J. Finkelstein, *The Sultantepe Tablets, Volume I* (London: British Institute of Archaeology at Ankara, 1957).

Oliver R. Gurney and Peter Hulin, *The Sultantepe Tablets, Volume II* (London: British Institute of Archaeology at Ankara, 1964).

Johannes Hackl, "Language Death and Dying Reconsidered: The Role of Late Babylonian as a Vernacular Language," Imperium and Officium Working Papers, 2011. Vienna: http://iowp.univie.ac.at/.

Johannes Hackl, "Materialien zur Urkundenlehre und Archivkunde der spätzeitlichen Texte aus Nordbabylonien" (PhD diss., Vienna University, 2013).

Johannes Hackl, "New Additions to the Rahimesu Archive: Parthian Texts from the British Museum and the World Museum Liverpool," in *Silver, Money and Credit: A Tribute to Robartus J. van der Spek on the Occasion of his 65th Birthday*, ed. Kristin Kleber and Reinhard Pirngruber (Leiden: Nederlands Instituut voor het Nabije Oosten, 2016), 87–106.

Johannes Haubold, Giovanni Lanfranchi, Robert Rollinger, and John Steele, *The World of Berossos* (Wiesbaden: Harrassowitz, 2013).

Hermann Hunger, *Babylonische und Assyrische Kolophone* (Kevelaer: Butzon & Bercker; Neukirchen-Vluyn: Neukirchener, 1968).

Hermann Hunger, *Astrological Reports to Assyrian Kings*, State Archives of Assyria 8 (Helsinki: Helsinki University Press, 1992).

Cynthia Jean, *La magie néo-assyrienne en contexte: Recherches sur le métier d'exorciste et le concept d'āšipūtu* (Helsinki: The Neo-Assyrian Text Corpus Project, 2006).

Alexander Jones, "Transmission of Babylonian Astronomy to Other Cultures," in *Handbook of Archaeoastronomy and Ethnoastronomy*, ed. Clive N. Ruggles (New York: Springer, 2015), 1877–81.

Michael Jursa, *Neo-Babylonian Legal and Administrative Documents: Typology, Contents and Archives* (Münster: Ugarit, 2005).

Michael Jursa, "Money-based Exchange and Redistribution: The Transformation of the Institutional Economy in First-millennium Babylonia," in *Autour de Polanyi: Vocabulaires, théories et modalités des échanges*, ed. Philippe Clancier, Francis Joannès, Pierre Rouillard, and Aline Tenu (Paris: de Boccard, 2005), 171–86.

Michael Jursa, "Die Söhne Kudurrus und die Herkunft der Neubabylonischen Dynastie," *Revue d'Assyriologie* 101 (2007): 125–36.

Michael Jursa, "The Transition of Babylonia from the Neo-Babylonian Empire to Achaemenid Rule," in *Regime Change in the Ancient Near East and Egypt: from Sargon of Agade to Saddam Hussein*, ed. Harriet Crawford (Oxford: Oxford University Press, 2007), 73–94.

Michael Jursa, "Taxation and Service Obligations in Babylonia from Nebuchadnezzar to Darius and the Evidence for Darius' Tax Reform," in *Herodot und das Persische Weltreich/ Herodotus and the Persian Empire*, ed. Robert Rollinger, Brigitte Truschnegg, and Reinhold Bichler (Wiesbaden: Harrassowitz, 2011), 431–48.

Michael Jursa, "Labor in Babylonia in the First Millennium BC," in *Labor in the Ancient World*, ed. Piotr Steinkeller and Michael Hudson (Dresden: ISLET, 2015), 345–96.

Wilfred G. Lambert, "A Late Assyrian Catalogue of Literary and Scholarly Texts," in *Kramer Anniversary Volume: Cuneiform Studies in Honor of Samuel Noah Kramer*, ed. Barry L. Eichler, Jane W. Heimerdinger, and Åke W. Sjöberg (Kevelaer: Butzon & Bercker; Neukirchen-Vluyn: Neukirchener, 1976), 313–18.

Wilfred G. Lambert, *Babylonian Oracle Questions* (Winona Lake: Eisenbrauns, 2007).

Alan Lenzi, *Secrecy and the Gods: Secret Knowledge in Ancient Mesopotamia and Biblical Israel* (Helsinki: The Neo-Assyrian Text Corpus Project, 2008).

Alan Lenzi, "The Uruk List of Kings and Sages and Late Mesopotamian Scholarship," *Journal of Ancient Near Eastern Religions* 8 (2008): 137–69.

Stephen Lieberman, "A Mesopotamian Background for the So-called Aggadic 'Measures' of Biblical Hermeneutics?" *Hebrew Union College Annual* 58 (1987): 157–225.

Stephen Lieberman, "Canonical and Official Cuneiform Texts: Towards an Understanding of Assurbanipal's Personal Tablet Collection," in *Lingering over Words: Studies in Ancient Near Eastern Literature in Honor of William L. Moran*, ed. Tzvi Abusch, John Huehnergard, and Piotr Steinkeller (Atlanta: Scholars Press, 1990), 305–36.

Alasdair Livingstone, *Mystical and Mythological Explanatory Works of Assyrian and Babylonian Scholars* (Oxford: Clarendon, 1986).

Alasdair Livingstone, "On the Organized Release of Doves to Secure Compliance of a Higher Authority," in *Wisdom, Gods and Literature: Studies in Assyriology in Honour of W.G. Lambert*, ed. Andrew R. George and Irving L. Finkel (Winona Lake: Eisenbrauns, 2000), 375–88.

Alasdair Livingstone, "Ashurbanipal: Literate or Not?" *Zeitschrift für Assyriologie* 97 (2007): 98–118.

Seton Lloyd and Nuri Gokçe, "Sultantepe: Anglo-Turkish Joint Excavations, 1952," *Anatolian Studies* 3 (1953): 27–47.

Stefan M. Maul, *Zukunftsbewältigung: Eine Untersuchung altorientalischen Denkens anhand der babylonisch-assyrischen Löserituale (Namburbi)* (Mainz: von Zabern, 1994).

Stefan M. Maul, "Die Tontafelbibliothek aus dem sogenannten »Haus des Beschwörungspriesters,«" *Assur-Forschungen: Arbeiten aus der Forschungsstelle »Edition Literarische Keilschrifttexte aus Assur« der Heidelberger Akademie der Wissenschaften*, ed. Stefan M. Maul and Nils P. Heeßel (Wiesbaden: Harrassowitz, 2010), 189–228.

Gilbert J. P. McEwan, "Arsacid Temple Records," *Iraq* 43 (1981): 131–43.

Gilbert J. P. McEwan, *Priest and Temple in Hellenistic Babylonia* (Wiesbaden: Steiner, 1981).

Otto Neugebauer, *Astronomical Cuneiform Texts, Volumes I–III* (Berlin: Springer, 1955).

Joachim Oelsner, "Von Iqīšâ und einigen anderen spätgeborenen Babyloniern," in *Studi su vicino Oriente antico dedicati alla memoria di Luigi Cagni*, ed. Simonetta Graziani (Napoli: Istituto Universitario Orientale, 2000), 797–813.

Mathieu Ossendrijver, *Babylonian Mathematical Astronomy: Procedure Texts* (New York: Springer, 2012).

Simo Parpola, "Assyrian Library Records," *Journal of Near Eastern Studies* 42 (1983): 1–29.

Simo Parpola, *Letters from Assyrian Scholars to the Kings Esarhaddon and Assurbanipal, Part II: Commentary and Appendices* (Kevelaer: Butzon & Bercker, 1983; repr., Winona Lake: Eisenbrauns, 2007).

Simo Parpola, *Letters from Assyrian and Babylonian Scholars*, State Archives of Assyria 10 (Helsinki: Helsinki University Press, 1993).

Simo Parpola, "Sequence of Post-canonical Eponyms," in *The Prosopography of the Neo-Assyrian Empire, Volume 1/I: A*, ed. Karen Radner (Helsinki: The Neo-Assyrian Text Corpus Project, 1998), xviii–xx.

Theophilus G. Pinches, "A Babylonian Tablet Dated in the Reign of Aspasine," *Babylonian and Oriental Record* 4 (1896): 131–35.

Reinhart Pirngruber, "The Historical Sections of the Astronomical Diaries in Context: Developments in a Late Babylonian Scientific Text Corpus," *Iraq* 75 (2013): 197–210.

Karen Radner, "Gatekeepers and Lock Masters: The Control of Access in Assyrian Palaces," in *Your Praise is Sweet: A Memorial Volume for Jeremy Black from Students, Colleagues and Friends*, ed. Heather D. Baker, Eleanor Robson, and Gábor Zólyomi (London: British Institute for the Study of Iraq, 2010), 269–80.

Julian E. Reade, "Assyrian Eponyms, Kings and Pretenders, 648–605," *Orientalia* 67 (1998): 255–65.

Julian E. Reade, "Ninive (Nineveh)," in *Reallexikon der Assyriologie und Vorderasiatischen Archäologie, Vol. 9*, ed. Dietz O. Edzard (Berlin: De Gruyter, 2001), 388–433.

George Reisner, *Sumerische-Babylonische Hymnen nach Thontafeln Griechischer Zeit* (Berlin: Spemann, 1896).

Eleanor Robson, *Mathematics in Ancient Iraq: A Social History* (Princeton: Princeton University Press, 2008).

Eleanor Robson, "The Clay Tablet Book in Sumer, Assyria and Babylonia," in *A Companion to the History of the Book*, ed. Simon Eliot and Jonathan Rose (Oxford: Blackwell, 2010), 67–83.

Eleanor Robson, "Empirical Scholarship in the Neo-Assyrian Court," in *The Empirical Dimension of Ancient Near Eastern Studies*, ed. Gebhardt Selz and Klaus Wagensonner (Vienna: LIT, 2011), 603–30.

Eleanor Robson, "The Production and Dissemination of Scholarly Knowledge," in *The Oxford Handbook of Cuneiform Culture*, ed. Karen Radner and Eleanor Robson (Oxford: Oxford University Press, 2011), 557–76.

Eleanor Robson, "Reading the Libraries of Assyria and Babylonia," in *Ancient Libraries*, ed. Jason König, Katerina Oikonomopoulos, and Greg Woolf (Cambridge: Cambridge University Press, 2013), 38–56.

Eleanor Robson, "Tracing Networks of Cuneiform Scholarship with Oracc, GKAB and Google Earth," in *Archaeologies of Text: Archaeology, Technology and Ethics*, ed. Matthew Rutz and Morag Kersel (Oxford: Oxbow Books, 2014), 142–63.

Eleanor Robson, "The Socio-economics of Cuneiform Scholarship after the 'End of Archives': Views from Borsippa and Uruk," in *At the Dawn of History: Ancient Near Eastern Studies in Honour of J. N. Postgate*, ed. Yagmur Heffron, Adam Stone, and Martin Worthington (Winona Lake: Eisenbrauns, 2017), 455–70.

Eleanor Robson, *Ancient Knowledge Networks: A Social Geography of Cuneiform Scholarship in the First Millennium BC* (forthcoming).

Eleanor Robson, "Who Wrote the Babylonian Astronomical Diaries?" in *Keeping Watch in Babylon: from Evidence to Text in the Astronomical Diaries*, ed. Johannes Haubold, John Steele, and Kathryn Stevens (Boston: Brill, forthcoming).

Eleanor Robson and Greta Van Buylaere, "Assyrian-Babylonian Scholarly Literacies" (unpublished manuscript).

Eleanor Robson and Kathryn Stevens, "Scholarly Tablet Collections in First-Millennium Assyria and Babylonia," in *The Earliest Libraries: Library Tradition in the Ancient Near East*, ed. Gojko Barjamovic and Kim Ryholt (Oxford: Oxford University Press, forthcoming).

Abraham J. Sachs and Hermann Hunger, *Astronomical Diaries and Related Texts from Babylonia, Volume III: Diaries from 164 B.C. to 61 B.C.* (Vienna: Austrian Academy of Sciences, 1996).

Daisuke Shibata, "A Nimrud Manuscript of the Fourth Tablet of the Series *Mīs pî*, CTN IV 170(+)188, and a *Kiutu* Incantation to the Sun God," *Iraq* 70 (2008): 189–203.

Ira Spar and Wilfred G. Lambert, eds., *Literary and Scholastic Texts of the First Millennium BC*, Cuneiform Texts in the Metropolitan Museum of Art 2 (New York: The Metropolitan Museum of Art, 2005).

Robartus J. van der Spek, "The Babylonian Temple during the Macedonian and Parthian Domination," *Bibliotheca Orientalis* 42 (1985): 541–62.
Robartus J. van der Spek, "The Astronomical Diaries as a Source for Achaemenid and Seleucid History," *Bibliotheca Orientalis* 50 (1993): 91–101.
Robartus J. van der Spek, "Cuneiform Documents on Parthian History: The Rahimesu Archive – Materials for the Standard of Living," in *Das Partherreich und seine Zeugnisse/The Arsacid Empire: Sources and Documentation*, ed. Josef Wiesehöfer (Stuttgart: Steiner, 1998), 205–58.
Kathryn Stevens, "Secrets in the Library: Protected Knowledge and Professional Identity in Late Babylonian Uruk," *Iraq* 75 (2013): 211–53.
Greta Van Buylaere, "A Palaeographic Analysis of Neo-Assyrian" (PhD diss., University of Udine, 2009).
Niek Veldhuis, "Levels of Literacy," in *The Oxford Handbook of Cuneiform Culture*, ed. Karen Radner and Eleanor Robson (Oxford: Oxford University Press, 2011), 68–89.
Pierre Villard, "L'education d'Assurbanipal," *Ktema* 22 (1997): 135–49.
Caroline Waerzeggers, "The Babylonian Revolts against Xerxes and the 'End of Archives,'" *Archiv für Orientforschung* 50 (2003/04): 150–73.
Caroline Waerzeggers, "The Babylonian Priesthood in the Long Sixth Century BC," *Bulletin of the Institute of Classical Studies* 54 (2011): 59–70.
Ernst Weidner, "Ein Kommentar zu den Schlangen-Omina," *Archiv für Orientforschung* 21 (1966): 46.
Ernst von Weiher, *Spätbabylonische Texte aus Uruk, 5te Band*, Ausgrabungen der Deutschen Forschungsgemeinschaft in Uruk-Warka, Endberichte 13 (Mainz: von Zabern, 1998).
Donald J. Wiseman, "Assyrian Writing Boards," *Iraq* 17 (1955): 3–13.
Donald J. Wiseman and Jeremy A. Black, *Literary Texts from the Temple of Nabû*, Cuneiform Texts from Nimrud 4 (London: British School of Archaeology in Iraq, 1996).

Mladen Popović

2 Multilingualism, Multiscripturalism, and Knowledge Transfer in the Dead Sea Scrolls and Graeco-Roman Judaea

2.1 Introduction

The Dead Sea Scrolls – about one thousand reconstructed manuscripts found in eleven caves between 1947–1956 that date to the third century BCE until the first century CE – provide a unique vantage point to study multilingualism, "multiscripturalism," and knowledge transfer. These three aspects offer a valuable entry into some of the cultural encounters in which people in ancient Judea took part.

The scrolls have been a treasure trove for all sorts of literary investigations into early Jewish and Christian traditions and thought-worlds, serving as a hub from which connections with diverse bodies of literary evidence from various geographic origins and different time-periods have been made. Much research also has been devoted to the social matrix of the presumed community or sect behind the scrolls, privileging certain textual evidence over others, for example, the so-called sectarian texts. This has become more difficult with the publication of all the scrolls material, questioning whether all manuscripts should be understood as one collection and attributed to one movement or community at a specific place. Since the early days of scrolls research, most scholars approached the manuscript finds as belonging to a distinct ancient Jewish group – the Essenes or the Qumran Community – inhabiting the site of Qumran. But more recently, this "single community at a single place"-framework has been questioned, and rightly so. Historical, literary, and religious studies analyses of the scrolls' contents indicate heterogeneity and religious diversity within the collection of texts on different levels. Literary heterogeneity and religious diversity have been related to

---

Note: The finalization for publication of this paper was carried out within the framework of the ERC Starting Grant of the European Research Council (EU Horizon 2020): The Hands that Wrote the Bible: Digital Palaeography and Scribal Culture of the Dead Sea Scrolls (HandsandBible #640497). Previous versions of this paper were presented at the conference *Multilingualism and the Transfer of Knowledge in Antiquity*, organized by Hindy Najman, at Yale University, 7–9 December 2014, and at the conference in Groningen April 2015 of which this volume represents the proceedings. I thank the participants at those conferences for their feedback. I also thank Barry Hartog, Gemma Hayes, and Eibert Tigchelaar for their suggestions and discussion when finalizing the paper for publication.

different models of communities behind the manuscripts, in terms of diverse but related communities at various localities that were behind these texts. Thus, for example, scholars argue that the different, conflicting versions of the sectarian Rule of the Community (Serekh ha-Yaḥad) were developed in Yaḥad communities that were geographically, not chronologically, distinct. Jerusalem, for example, may have been one such location outside Qumran. This analysis of the Serekh manuscripts is then extrapolated to the Qumran collection as a whole.

The texts in the different caves attest to various scribal practices, among which a so-called Qumran scribal practice,[1] to multilingualism through the use of languages such as Hebrew, Aramaic, and Greek, and to "multiscripturalism" through the use of various scripts, sometimes in the same manuscript, such as square script for Hebrew and Aramaic, palaeo-Hebrew, Greek, and Cryptic scripts. These sociolinguistic and scribal features are also significant in light of proposals that consider the choice of Hebrew as an anti-language or holy language to reflect the social context of the movement behind these texts having been one of isolation.[2] Sociolinguistic method and theory of multilingualism and language ideology[3] may present us with alternative models that better explain the hetero-

---

[1] Emanuel Tov, *Scribal Practices and Approaches Reflected in the Texts Found in the Judean Desert*, Studies on the Texts of the Desert of Judah 54 (Leiden: Brill, 2004); idem, "Scribal Practices and Approaches Revisited," *Hebrew Bible and Ancient Israel* 3 (2014): 355–67; Eibert J. C. Tigchelaar, "Assessing Emanuel Tov's 'Qumran Scribal Practice,'" in *The Dead Sea Scrolls: Transmission of Traditions and Production of Texts*, ed. Sarianna Metso, Hindy Najman, and Eileen Schuller, Studies on the Texts of the Desert of Judah 92 (Leiden: Brill, 2010), 173–205.

[2] William H. Schniedewind, "Qumran Hebrew as an Antilanguage," *Journal of Biblical Literature* 118 (1999): 235–52; idem, "Linguistic Ideology in Qumran Hebrew," in *Diggers at the Well: Proceedings of a Third International Symposium on the Hebrew of the Dead Sea Scrolls and Ben Sira*, ed. Takamitsu Muraoka and John F. Elwolde, Studies on the Texts of the Desert of Judah 36 (Leiden: Brill, 2000), 245–55; idem, *A Social History of Hebrew: Its Origins Through the Rabbinic Period* (New Haven: Yale University Press, 2013); Steven Weitzman, "Why Did the Qumran Community Write in Hebrew?" *Journal of the American Oriental Society* 119 (1999): 35–45; Gary A. Rendsburg, "Qumran Hebrew (With a Trial Cut [1QS])," in *The Dead Sea Scrolls at 60: Scholarly Contributions of New York University Faculty and Alumni*, ed. Lawrence H. Schiffman and Shani Tzoref, Studies on the Texts of the Desert of Judah 89 (Leiden: Brill, 2010), 217–46.

[3] See, e.g., Kormi Anipa, "The Use of Literary Sources in Historical Sociolinguistic Research," in *The Handbook of Historical Sociolinguistics*, ed. Juan Manuel Hernández-Campoy and Juan Camilo Conde-Silvestre (Oxford: Wiley-Blackwell, 2012), 170–90; Hanna Rutkowska and Paul Rössler, "Orthographic Variables," in *The Handbook of Historical Sociolinguistics*, ed. Juan Manuel Hernández-Campoy and Juan Camilo Conde-Silvestre (Oxford: Wiley-Blackwell, 2012), 213–36; Herbert Schendl, "Multilingualism, Code-switching, and Language Contact in Historical Sociolinguistics," in *The Handbook of Historical Sociolinguistics*, ed. Juan Manuel Hernández-Campoy and Juan Camilo Conde-Silvestre (Oxford: Wiley-Blackwell, 2012), 520–33; Florian Coulmas, *Writing and Society: An Introduction* (Cambridge: Cambridge University Press, 2012);

geneous collections of writings from the Judaean Desert, which strongly suggest multiple standards.

Bearing in mind the rich diachronic and multifaceted insights all this data and research have given us, I wish to redirect the focus on the people behind the scrolls again, not in the sense of a single community at a single place, but to understand the collections of manuscripts as a reflection of a textual community, understood as a micro-society in antiquity organized around a common understanding of texts.[4] However one conceives of the configuration of the people behind the scrolls, texts were central in their social activities. The wealth of texts attests that people were occupied with the interpretation of and commentary on scripture, legal issues and community building, but also with science, magic and the writing of history. These people were not isolated but participated in various ways in ancient Mediterranean intellectual networks.[5] Through the writing, copying, and studying of texts, the scrolls' anonymous scribes and teachers constructed a textual community of a highly intellectual and scholarly character.[6] The textual community behind the Dead Sea Scrolls was not only an ancient Judean phenomenon but also an ancient Mediterranean phenomenon. Taking multilingualism, multiscripturalism, and knowledge transfer as key issues will provide us with an entry into this ancient Mediterranean textual community and also show its entangled history with other intellectual and scholarly communities, both near and far.

Robert Bayley, Richard Cameron, and Ceil Lucas, eds., *The Oxford Handbook of Sociolinguistics* (Oxford: Oxford University Press, 2013); Sari Pietikäinen and Helen Kelly-Holmes, eds., *Multilingualism and the Periphery* (Oxford: Oxford University Press, 2013).

[4] See, e.g., Brian Stock, *The Implications of Literacy: Written Language and Models of Interpretation in the Eleventh and Twelfth Centuries* (Princeton: Princeton University Press, 1983); idem, *Listening for the Text: On the Uses of the Past* (Baltimore: Johns Hopkins University Press, 1990).

[5] See, e.g., Mladen Popović, *Reading the Human Body: Physiognomics and Astrology in the Dead Sea Scrolls and Hellenistic-Early Roman Period Judaism*, Studies on the Texts of the Desert of Judah 67 (Leiden: Brill, 2007); Jonathan Ben-Dov, *Head of All Years: Astronomy and Calendars at Qumran in their Ancient Context* (Leiden: Brill, 2008); Jonathan Ben-Dov and Seth L. Sanders, eds., *Ancient Jewish Sciences and the History of Knowledge in Second Temple Literature* (New York: New York University Press, 2014); Pieter B. Hartog, *Pesher and Hypomnema: A Comparison of Two Commentary Collections from the Hellenistic-Roman Period*, Studies on the Texts of the Desert of Judah 121 (Leiden: Brill, 2017). See also Mladen Popović, Myles Schoonover, and Marijn Vandenberghe, eds., *Jewish Cultural Encounters in the Ancient Mediterranean and Near Eastern World*, Supplements to the Journal for the Study of Judaism 178 (Leiden: Brill, 2017).

[6] Mladen Popović, "Qumran as Scroll Storehouse in Times of Crisis? A Comparative Perspective on Judaean Desert Manuscript Collections," *Journal for the Study of Judaism* 43 (2012): 551–94.

2.2 Multilingualism

When dealing with the multiple languages and scripts in the Dead Sea Scrolls in relation to actual language use and proficiency of the people behind the manuscripts an important presupposition is often in operation to frame the linguistic evidence: namely, that we are dealing with a small, isolated, marginal (and even weak) community at the site of Khirbet Qumran.

For example, Steven Weitzman (following Chaim Rabin and Bernard Spolsky) has argued for a special status of the Hebrew language, while William Schniedewind has argued for a special form of Hebrew, so-called Qumran Hebrew, over against what is perceived as the vernacular Aramaic and Hebrew, Mishnaic Hebrew, of the time. Weitzman has asked why the presumed Qumran community wrote in Hebrew, whereas Schiedewind asks why they wrote a specific form of Hebrew.

Weitzman, as other scholars, points to Jub. 12:25–27 and 4Q464. The book of Jubilees refers to Hebrew as the language of creation and 4Q464 speaks of the holy language, *lishon ha-qodesh*, most probably referring to Hebrew.[7] Weitzman refers to an article by Spolsky,[8] presuming that first-century Jews in their multilingual environment tended to use the language that asserted the most advantageous social membership for them in the proposed interaction. In other words, Weitzman assumes social advantages of using Hebrew. More specifically, he suggests Hebrew may have been perceived as "the linguistic prerequisite for membership in a supernatural community, either the community at the End of Days or that of the angels in the heavenly temple."[9] In a multilingual environment, Weitzman sees the use of Hebrew and the avoidance of other "mundane" languages as a linguistic ideology signalling an identity of these people apart from others.

Although Weitzman asks why they *wrote* in Hebrew, his sociological explanation seems to imply more than merely writing in Hebrew. This seems a very idealized view of the people behind the Dead Sea Scrolls and their textual production and ignores other evidence that does not assert the exclusive use of Hebrew. First, it seems that the scribe of the Great Isaiah scroll from Qumran Cave 1 was an Aramaic speaker, or at least influenced by the Aramaic language.[10] More

---

[7] See now Willem F. Smelik, "Holy Tongue," in *Rabbis, Language and Translation in Late Antiquity* (Cambridge: Cambridge University Press, 2013), 42–99.
[8] Bernard Spolsky, "Jewish Multilingualism in the First Century: An Essay in Historical Sociolinguistics," in *Readings in the Sociology of Jewish Languages*, ed. Joshua A. Fishman (Leiden: Brill, 1985), 34–50.
[9] Weitzman, "Qumran Community Write in Hebrew," 45.
[10] Edward Y. Kutscher, *The Language and Linguistic Background of the Isaiah Scroll (1QIsaᵃ)*, Studies on the Texts of the Desert of Judah 6 (Leiden: Brill, 1974); Martin G. Abegg, "Linguistic

generally, "[t]he amount of Aramaic influence in the Hebrew Qumran scrolls can best be explained as reflecting the bilingualism of the authors and their readers."[11] Second, the citation of two versions of Hab 2:16 has been adduced as evidence that the writer of the Pesher Habakkuk used a Greek manuscript in addition to a Hebrew one (see also Hab 1:17).[12] I am not arguing that Hebrew was not important, but these two examples show that in our understanding of the scribal process of text production the assumption of monolingual prejudice or preference does not do justice to the variegated evidence, which points to multilingual competencies.

While Weitzman is operating with a notion of the perception of Hebrew in ancient Judaism that is more widespread and in general correct but not in explaining why they wrote in Hebrew, Schniedewind works with a more specific idea of a particular form of Hebrew. This idea he set out in two earlier articles and repeats in his recent book, *A Social History of Hebrew*.[13] Schniedewind perceives the use of code and symbolic terminology, archaisms or pseudoclassicizing tendencies, an avoidance of Aramaic, and also elements of Emanuel Tov's Qumran scribal practice (such as long pronominal forms, both independent and suffixed; suffixed *ah* in a variety of adverbials; long forms of the first-person imperfect; writing of the divine name in palaeo-Hebrew), as indicators of the community's language ideology.[14]

The linguistic data Schniedewind uses is not up to date and also does not support the notion of anti-language that he introduces for Qumran Hebrew.[15] To give one example: concrete data relating to the production and use of specific manuscripts is ignored, such as the *tefillin* of which more than half consistently use the long forms, which may speak against the presumed artificiality of Qumran Hebrew. Of the *tefillin* one may ask: "Is this because their scribes wanted

---

Profiles of the Isaiah Scrolls," in *Qumran Cave 1.II, The Isaiah Scrolls, Part 2: Introductions, Commentary, and Textual Variants*, Eugene Ulrich and Peter W. Flint, Discoveries in the Judaean Desert 32 (Oxford: Clarendon, 2010), 25–41, at 41.

11 Jan Joosten, "Hebrew, Aramaic, and Greek in the Qumran Scrolls," in *The Oxford Handbook of the Dead Sea Scrolls*, ed. Timothy H. Lim and John J. Collins (Oxford: Oxford University Press, 2010), 351–74, at 359.

12 Timothy H. Lim, "The Qumran Scrolls, Multilingualism, and Biblical Interpretation," in *Religion in the Dead Sea Scrolls*, ed. John J. Collins and Robert A. Kugler (Grand Rapids: Michigan, 2000), 57–73, at 70–72. Cf., however, Hartog, *Pesher and Hypomnema*, 154–58.

13 See note 3 above.

14 Schniedewind, *Social History of Hebrew*, 173–89.

15 As Eibert Tigchelaar has discussed at the 2013 International Organisation for Qumran Studies meeting in Munich, see "Sociolinguistics and Which Dead Sea Scrolls," forthcoming in the conference proceedings (for now see Tigchelaar's academia.edu site). See also Weitzman, "Qumran Community Write in Hebrew," 37.

to make these biblical texts even more archaic than they already were? Or perhaps because their scribes, who probably wrote these texts from memory, were not constrained by the graphic conventions of written *Vorlages*?"[16] The quantification of linguistic data and the correlation of data sets in the scrolls should be matched by an assessment of scribal production and profiling that is based on the empirical traces of scribal activity. Furthermore, Schniedewind seems undecided in his 2013 book, *A Social History of Hebrew*, in characterizing the people behind the Dead Sea Scrolls. On the one hand, he repeats from his earlier articles a more traditional framing of the Qumran community as a small, isolated community, while, on the other hand, he refers to more recent scholarship that posits multiple communities behind these texts, without fully integrating such more recent trends and drawing clear conclusions for what this means for the relationship between specific texts and the social reality behind them, especially from his sociolinguistic perspective on a specific language form being and anti-language.[17]

In light of the evidence that is now available, the notion of anti-language is not useful to understand the linguistic evidence from the scrolls. There is no basis to see Qumran Hebrew as intentionally set apart from Hebrew used elsewhere in Judaea at the time. Instead, we should consider approaching the heterogeneous material from the perspective of multiple standards.[18] The material from Qumran is linguistically heterogeneous, not just because of multiple languages such as Hebrew, Aramaic, and Greek, but also for example with regard to orthography and morphology in such a way that consistency does not appear (which Schniedewind also acknowledges with regard to Tov's Qumran scribal practice). The point is that we need not reckon with linguistic consistency but allow for multiple standards to understand the evidence in a more complex context than that of a presumed small, isolated, and marginal group at Qumran. I suggest that the manuscripts from the caves near Qumran and what they represent should be no longer framed as centre-periphery in the sense that Qumran was deviating from a standard norm.

---

16 Tigchelaar, "Sociolinguistics."
17 Compare Schniedewind, *Social History of Hebrew*, 177, 178 ("This small, isolated religious community on the north shore of the Dead Sea used language ideologically as a means of differentiating and further insulating themselves…. Small, weak, and marginal religious communities such as the *yaḥad* community typically cultivate linguistic idiosyncrasies in order to enhance group identity."), and 180 (but see also 173–74), commenting on the inconsistent implementation of the Qumran scribal practice ("the realization that the sectarian scrolls were copied by a variety of *yaḥad* scribes in a variety of places over a two-hundred year period accounts for the inconsistencies in sectarian orthography. Indeed, the lack of complete standardization points to a loose social structure of the group….").
18 Tigchelaar, "Sociolinguistics."

Recent sociolinguistic research on multilingualism and minority languages from a centre-periphery dynamics perspective may be useful to reframe our approach to the Dead Sea Scrolls in relation to our presuppositions of the broader linguistic situation in ancient Judaea in the Greco-Roman period.[19]

The notions of "centre" and "periphery" are not given, but should instead, Sari Pietikäinen and Helen Kelly-Holmes argue, be understood as discursive constructs, products of social interaction, reflecting the circumstances and dynamics of their construction. Moreover, centre-periphery approaches also allow for the possibility for peripheral sites to become centres of normativity rather than places to which norms are disseminated. While the centre has traditionally been seen as the source of norms to be adopted in peripheries, the dynamics of the centre–periphery relationship might instead lead to the derivation of new and multiple normativities.[20] This is important in relation to the heterogeneous character of the evidence from Qumran, allowing for the perspective of multiple standards instead of one standard.

With regard to linguistic evidence, Pietikäinen and Kelly-Holmes identify

---

[19] In addition to the references in note 4 above, the scholarly literature on multilingualism, bilingualism, diglossia, and code-switching in the ancient world and ancient Judaea specifically is fast-growing. For orientation, see, e.g., James N. Adams, Mark Janse, and Simon Swain, eds., *Bilingualism in Ancient Society: Language Contact and the Written Text* (Oxford: Oxford University Press, 2002); James N. Adams, *Bilingualism and the Latin Language* (Cambridge: Cambridge University Press, 2003); Willem Smelik, "Code-switching: The Public Reading of the Bible in Hebrew, Aramaic and Greek," in *Was ist ein Text? Alttestamentliche, ägyptologische und altorientalische Perspektiven*, ed. Ludwig Morenz and Stefan Schorch (Berlin: De Gruyter, 2007), 123–47; Hannah M. Cotton, Robert G. Hoyland, Jonathan J. Price, and David J. Wasserstein, eds., *From Hellenism to Islam: Cultural and Linguistic Change in the Roman Near East* (Cambridge: Cambridge University Press, 2009); Dorothy J. Thompson, "The Multilingual Environment of Persian and Ptolemaic Egypt: Egyptian, Aramaic, and Greek Documentation," in *The Oxford Handbook of Papyrology*, ed. Roger S. Bagnall (Oxford: Oxford University Press, 2009), 395–417; Willem Smelik, "The Languages of Roman Palestine," in *The Oxford Handbook of Jewish Daily Life in Roman Palestine*, ed. Catherine Hezser (Oxford: Oxford University Press, 2010), 122–41; Alex Mullen and Patrick James, eds., *Multilingualism in the Graeco-Roman Worlds* (Cambridge: Cambridge University Press, 2012); Steven D. Fraade, "Language Mix and Multilingualism in Ancient Palestine: Literary and Inscriptional Evidence," *Jewish Studies* 48 (2012): 1–40; Alex Mullen, *Southern Gaul and the Mediterranean: Multilingualism and Multiple Identities in the Iron Age and Roman Periods* (Cambridge: Cambridge University Press, 2013); Michael O. Wise, *Language and Literacy in Roman Judaea: A Study of the Bar Kokhba Documents* (New Haven: Yale University Press, 2015).

[20] Sari Pietikäinen and Helen Kelly-Holmes, "Multilingualism and the Periphery," in *Multilingualism and the Periphery*, ed. Sari Pietikäinen and Helen Kelly-Holmes (Oxford: Oxford University Press, 2013), 1–16.

at least two language ideological formations that have structured our understanding of multilingualism and consequently have had an influence on how individuals experience "languages" and talk about them. One powerful conceptualization ... has been the idea that languages are autonomous and unified entities ... with an "essential" or natural relationship with a particular territory or the collective identity of a particular group, and essentially "different" and "separate" from each other. ... At the same time, [they] have also documented an alternative ideological formation – that manifests itself, for example, in discourses of plurilingual identities and competencies or "polycentric" and "polynomic" languages and language practices. ... It can be argued that this perspective also captures the experiences of many multilingual speakers more appropriately by recognizing the inherent diversity and hybridity that characterizes multilingual living.[21]

These insights from a centre-periphery dynamics perspective may help us in reframing our ideas about the actual use of Aramaic, Hebrew, and Greek and to suggest plurilingual identities and competencies for at least some of the people behind the scrolls, as the examples discussed above of a possible Aramaic speaker that produced the great Isaiah Scroll in Hebrew and the possible use of a Greek *Vorlage* for the Pesher Habakkuk, also in Hebrew, may indicate. And then there is also the reference to the community official of the overseer, the so-called Mebaqqer, who is expected to know every language (Damascus Document 14:8–10; 4Q266 10 i 3), which may indicate an expected plurilingual competency from such an official precisely because of the group(s) that person was overseeing being characterized by plurilingual identities and competencies.[22] Insights from a centre-periphery dynamics perspective may aid us so as not to fossilize a new construct of directionality but to broaden our approach to the linguistic and literary landscape of ancient Judaea that accounts for the evidence in a differentiated manner. This may modify how we perceive Jerusalem as a centre for the production and transmission of texts and traditions vis-à-vis other parts in ancient Judea as well as

---

21 Pietikäinen and Kelly-Holmes, "Multilingualism and the Periphery," 8–9.
22 See also below on 4Q477. This passage from the Damascus Document and its possible implications for language competence has not received much attention in scholarship, in part perhaps because of the fragmentary manuscript evidence; see Martin Hengel, "Qumrân und der Hellenismus," in *Qumrân: Sa piété, sa théologie et son milieu*, ed. Matthias Delcor, Bibliotheca Ephemeridum Theologicarum Lovaniensium 46 (Paris: Duculot and Leuven: Leuven University Press, 1978), 333–72, at 340; G. Wilhelm Nebe, "Das Sprachvermögen des Mebaqqer in *Damaskusschrift* XIV, 10," *Revue de Qumrân* 16/62 (1993): 289–91; Weitzman, "Qumran Community Write in Hebrew," 35 n. 1; Marcus K. M. Tso, *Ethics in the Qumran Community: An Interdisciplinary Investigation*, Wissenschaftliche Untersuchungen zum Neuen Testament 292 (Tübingen: Mohr Siebeck, 2010), 125 n. 20. The relationship between the Damascus Document manuscripts and the Rule of the Community manuscripts as well as their connection with a community or communities behind the collections of scrolls are issues not dealt with here.

how we see ancient Judea as part of an ancient Mediterranean network of textual and intellectual communities engaged in knowledge transfer.

Furthermore, there is no need to isolate evidence on the basis of presumed language competencies. This has often been done for the evidence from Qumran Cave 7, where only Greek manuscripts were found. Cave 7 is often distinguished as the cave of a single inhabitant with a particular interest in Greek manuscripts.[23] However, this impression of exclusively Greek writings from Cave 7 is in need of some correction. There is an inscription of the name *Rom'a* in Aramaic characters that occurs twice on a large jar that was found in Cave 7 (7Q-Arch 2 heb/ar), and in one of his preliminary publications, Roland de Vaux refers to a small leather fragment in Hebrew from Cave 7, which was either a mistaken attribution or this fragment has since been overlooked. In most other Qumran caves, we find Aramaic texts alongside Hebrew and, of course, some Greek manuscripts were also found in Cave 4. This does not suggest a linguistic division within the collection or collections of scrolls. The presence of only Greek texts in Qumran Cave 7 should not be over-interpreted without other evidence of writing from this cave also being taken into account.[24] Instead, the evidence points to broader plurilingual competencies.

In addition to this centre-periphery dynamics perspective that stresses heterogeneity of linguistic practices, I would like to add another important observation concerning multilingualism in the ancient Mediterranean. James Clackson has argued against the suggestion of Ramsay MacMullen that "after the advent of Roman rule the local vernaculars were situated in socially or geographically isolated pockets of the Empire: the rural population of the countryside were largely monolingual in the local vernacular, but urban dwellers and upper classes were proficient in Latin and Greek."[25] Instead, for the Roman Near East and Egypt he argues that: "Rather than a monolingual countryside, with some bilingual speakers resident in towns and cities, it seems that there was stable bilingualism in the countryside, where local languages were used alongside Latin and Greek, and the bulk of the monolingual speakers were urban dwellers, proficient in Latin or Greek (or both) but often not in the local vernaculars."[26]

The impression of a bilingual or multilingual countryside in the Roman Near East is confirmed when we look at the Judaean Desert manuscript finds, taking

---

23 See recently, e.g., Wise, *Language and Literacy*, 325–26, 334.
24 Popović, "Qumran as Scroll Storehouse," 571.
25 James Clackson, "Language Maintenance and Language Shift in the Mediterranean World during the Roman Empire," in *Multilingualism in the Graeco-Roman Worlds*, ed. Alex Mullen and Patrick James (Cambridge: Cambridge University Press, 2012), 36–57, at 47.
26 Clackson, "Language Maintenance," 49.

this more broadly to include not only the Dead Sea Scrolls from the eleven caves near Qumran, but all manuscript finds in the desert area west of the Dead Sea. Personal archives that were left in Judaean Desert caves, such as the first to second-centuries CE Babatha and Salome Komaise archives from Naḥal Ḥever that have Greek next to Aramaic and Nabatean, show a multifaceted engagement with different languages in the different settings of everyday life, not only in urban centres but also in the countryside.[27] With regard to the Bar Kokhba letters there is the famous example of the letter in Greek (P.Yadin 52) in which the writer, Soumaios, apologizes for not having written it in Hebrew, which, scholars suggest, may have been expected from him.[28]

When considering the literary texts from Naḥal Ḥever, Wadi Murabbaʿat and also Masada it is clear that Hebrew was used in the countryside next to Aramaic, Nabatean and Greek. One might object, saying that the text finds from Naḥal Ḥever and Wadi Murabbaʿat date to the second century, but some of the literary texts, in Hebrew and Greek, are dated to the late first century BCE and early first century CE. These texts may have been in a family for several generations. Those who were unable to read them would still have had access to such literary texts: those who had attained a sufficient level of literary literacy would have read such literary manuscripts to those who could not read, perhaps in the social context of family or friends, or even in the larger social context of the village.[29]

---

[27] Catherine Hezser, *Jewish Literacy in Roman Palestine*, Texts and Studies in Ancient Judaism 81 (Tübingen: Mohr Siebeck, 2001), 309–19; Wise, *Language and Literacy*. Different legal systems are sometimes related to different languages; see Jacobine G. Oudshoorn, *The Relationship between Roman and Local Law in the Babatha and Salome Komaise Archives: General Analysis and Three Case Studies on the Law of Succession, Guardianship and Marriage*, Studies on the Texts of the Desert of Judah 69 (Leiden: Brill, 2007).
[28] Hezser, *Jewish Literacy*, 277–79; Wise, *Language and Literacy*, 245–51.
[29] Popović, "Qumran as Scroll Storehouse," 575; Mladen Popović, "Scribal Culture of the Hebrew Bible and the Burden of the Canon: Human Agency and Textual Production and Consumption in Ancient Judaism," in *Jeremiah's Scriptures: Production, Reception, Interaction, and Transformation*, ed. Hindy Najman and Konrad Schmid, Supplements to the Journal for the Study of Judaism 173 (Leiden: Brill, 2016), 253–58, at 257–58. Wise, *Language and Literacy*, 279–355 supports these inferences. Wise argues that Hebrew was the usual language of literature in multilingual Roman Judaea, not only at Qumran but also elsewhere. On the basis of his research into signature literacy as an indicator not only of writing but also of reading abilities, Wise suggests that during the first century BCE until the second century CE 65–80 per cent of Judaeans spoke a form of Hebrew (a vernacular termed proto-Mishnaic Hebrew). While Aramaic was the primary language of daily and documentary writing for ordinary people, Wise argues that Aramaic literary texts were the domain of elite intellectuals (Wise rules out Mas1p as a possibly Aramaic literary text from Masada, pp. 302, 327). As for Greek, Wise suggests that it was spoken to a considerable amount in Roman Judaea. He understands the Greek manuscripts from the Judaean

Not only was the Judaean countryside multilingual, but also "multi-literary" in the sense that high literary culture in Hebrew was not limited to urban centres but was also to a certain degree accessible in the countryside. Comparative analysis of the text finds in the Judaean Desert indicates the spread of literary texts within various strata of ancient Jewish society, outside of urban centres such as Jerusalem.[30] Michael Wise has argued that these strata were not limited to the top 1–2 per cent of society but that they should be understood as to include the top quartile percentage of the population. This does not mean that all those in the top quartile had mastered a sufficient level of literary literacy: this level Wise attributes to 5–10 per cent of Judaean men by one definition of literacy or up to 16 per cent by another definition. Different levels of literacy together with interdependency, often within the context of family, between literates of various levels and illiterates would have ensured a broader access to literary texts.[31]

The context, number of literary texts, and character of texts of the Judaean Desert manuscript finds reveal a differentiated engagement with literary texts by different kinds of people in Jewish society at the time. Members of the local rural elite indeed had access to some of their society's literary texts, but they did not engage with them in the same manner as, for example, someone such as Flavius Josephus or some of those behind the Dead Sea Scrolls. Many, if not most, of the literary literates among the rural, local elite had less time and money, and therefore leisure, to spend on studying their ancestral literary traditions. They mostly were positioned considerably farther down the social scale than those at the centre of power, such as Flavius Josephus, or those, such as some of the people behind the scrolls from the caves near Qumran, whose social infrastructure apparently supported an intensive and scholarly engagement with study of the ancestral traditions and other bodies of learned knowledge. The movement behind the scrolls can be characterized as a milieu of Jewish intellectuals or

---

Desert find sites to be examples of the phenomenon of alternative literacy: paralleling a Semitic track, there was an alternative educational path in Greek, up until the level of literary literacy. And then there were also tandem or dual literacies: those, presumably very few, who mastered both the Semitic and the Greek educational path, Flavius Josephus being a prime example of this phenomenon, but Wise also suggests a lesser well-known but fascinating example in the figures of Masabala b. Simon and his brothers.

30 Popović, "Qumran as Scroll Storehouse."
31 Wise, *Language and Literacy*, 40, 309–16, 344, 349–50. These calculations make more concrete earlier proposals for a smaller scale of dissemination in ancient Judaea, limited to leaders and their followers coming from the better off strata; see Albert I. Baumgarten, *The Flourishing of Jewish Sects in the Maccabean Era: An Interpretation*, Supplements to the Journal for the Study of Judaism 55 (Leiden: Brill, 1997), 127.

scholars who were engaged at a very high level with their ancestral traditions.[32] The Dead Sea Scrolls from the caves near Qumran attest to the vibrant and exciting presence in Graeco-Roman Judaea of a scholarly literacy that was connected with scholarly learning from elsewhere in the ancient Mediterranean and Near Eastern world.[33]

Insights from a centre-periphery dynamics perspective, that stresses heterogeneity of linguistic practices, need to be taken together with the manuscript evidence from the Judaean Desert as it attests multilingual competencies, congruent with other evidence from the Roman Near East that indicates a bilingual or multilingual countryside. All this calls for a more nuanced interpretation of multilingualism that cannot be neatly cut into isolated pockets of monolingual language ideology.[34]

2.3 Multiscripturalism

When it comes to the use of scripts in ancient Judaism, the more general observation seems often to be applied that "there existed in ancient times a strong bond between a language and its script."[35] Assuming that different languages tend to use different scripts, "when a second language is imposed on or taken up by a people, they may also acquire a second script .... Bilingualism thus interacts in interesting ways with biculturalism."[36] Ancient Judaism presents us with the interesting case that the Hebrew language remained in use but that sometime since the late sixth century BCE a switch was made to write that language in the Aramaic script (now referred to as the square script). The details for the

---

[32] Popović, "Qumran as Scroll Storehouse." Likewise, Wise, *Language and Literacy*, 327–31 characterizes the people behind the scrolls as hyperliterates. He emphasizes the presence of Aramaic literary texts as a key element for such a characterisation. Charlotte Hempel, *The Qumran Rule Texts in Context: Collected Studies*, Texts and Studies in Ancient Judaism 154 (Tübingen: Mohr Siebeck, 2013), 303–37 has argued that the Cave 4 manuscripts reflect a learned collection intended for an elite group within the movement.
[33] See the discussion further below.
[34] See Wise, *Language and Literacy*, 227, 243, 251 for a discussion about the preference under Bar Kokhba for Hebrew and use of the bookhand in letters.
[35] Joseph Naveh, *Early History of the Alphabet* (Jerusalem: Magnes and Leiden: Brill, 1982), 114; Willem F. Smelik, "*Ashurit* and Alphabet," in *Rabbis, Language and Translation in Late Antiquity* (Cambridge: Cambridge University Press, 2013), 271–322, at 275.
[36] James N. Adams and Simon Swain, "Introduction," in *Bilingualism in Ancient Society: Language Contact and the Written Text*, ed. James N. Adams, Mark Janse, and Simon Swain (Oxford: Oxford University Press, 2002), 1–20, at 5–6.

reasons why this shift occurred remain elusive. The influence of the international Aramaic culture on Judah at the time of the Persian Empire may have been an important factor in the change of script.[37] Despite this change in script, it is possible that both scripts, the older form of palaeo-Hebrew and the more recent form of Aramaic script, remained in use simultaneously since the Persian period to write the Hebrew language.[38] Even if there was an awareness that the Aramaic or square script and the Hebrew language had distinct histories, script and language were inextricably linked in the perception of their users.[39] While this may indeed apply to some if not most of the users in ancient Judaea, it is also important to recall another general observation that "script and language are not the same thing, and neither of them is an unambiguous marker of ethnic identity."[40]

A similar complexity as with multilingualism in ancient Judaea applies to the instances of multiscripturalism, not just in the scrolls from Qumran but also from elsewhere in the Judean Desert. By far, most manuscripts from the scrolls near Qumran were written in the Aramaic or square script, but the use of other scripts – palaeo-Hebrew, Greek, and Cryptic – is clearly attested. Explanations for the use of these various scripts have sometimes focused only on one script, but examples of manuscripts in which more than one script was used remind us that in practice the decision to use such scripts was not made in splendid isolation and that at least some people possessed "pluriscriptural" competencies.

Attesting to Greek language and script use are Greek literary manuscripts (and perhaps also a few documentary manuscripts)[41] that were found in Caves 4

---

[37] Smelik, "*Ashurit* and Alphabet," 275.
[38] Smelik, "*Ashurit* and Alphabet," 275–78. Another possibility is that the palaeo-Hebrew script was reintroduced during the Hasmonean period in the second century BCE. See also David Vanderhooft, "'el-mĕdînâ ûmĕdînâ kiktābāh: Scribes and Scripts in Yehud and in Achaemenid Transeuphratene," in *Judah and the Judeans in the Achaemenid Period: Negotiating Identity in an International Context*, ed. Oded Lipschits, Gary N. Knoppers, and Manfred Oeming (Winona Lakes: Eisenbrauns, 2011), 529–44, at 539; Eibert Tigchelaar, "The Material Variance of the Dead Sea Scrolls: On Texts and Artefacts," *HTS Teologiese Studies/Theological Studies* 72/4 (2016): 1–6, at 2–3.
[39] Smelik, "*Ashurit* and Alphabet," 271–73.
[40] Fergus Millar, "Introduction: Documentary Evidence, Social Realities and the History of Language," in *From Hellenism to Islam: Cultural and Linguistic Change in the Roman Near East*, ed. Hannah M. Cotton, Robert G. Hoyland, Jonathan J. Price, and David J. Wasserstein (Cambridge: Cambridge University Press, 2009), 1–12, at 6 (Millar refers to studies by Michael C. A. MacDonald). See also Mullen, *Southern Gaul and the Mediterranean*, 14: "No direct equation can be made between ethnicity, culture and language, though all three are deeply entwined."
[41] Only a few examples of documentary texts, such as accounts, lists of names and scribal practices, were found, although the provenance of a number of them from Qumran Cave 4 is disputed; see Hannah M. Cotton and Ada Yardeni, *Aramaic, Hebrew and Greek Documentary Texts from Naḥal Ḥever and Other Sites, with an Appendix Containing Alleged Qumran Texts*, Discoveries in

and 7 near Qumran, and Greek documentary texts, and also a few literary texts, that appeared at other Judaean Desert sites.[42] Recalling the discussion above about the plurilingual competencies of the functionary of the Mebaqqer and what this indicates about the plurilingual identities of the group(s) this functionary was overseeing, the text Rebukes Reported by the Overseer (4Q477) serves as a further indicator of such plurilingual identities. In addition to two people with Hebrew surnames, 4Q477 2 ii 5 lists the Greek epithet of one Ḥananiah Notos.[43] The Greek epithet here may indicate a deeper engagement with the Greek language and signal a bilingual identity or competency.[44]

The scrolls have also provided us with evidence for the use of palaeo-Hebrew script: fifteen manuscripts from Qumran were written entirely in palaeo-Hebrew, and an additional one comes from Masada. These are mainly copies of the books of Moses (Genesis to Deuteronomy) and Job.[45] Also, individual palaeo-Hebrew

---

the Judaean Desert 27 (Oxford: Clarendon, 2010), 6, 283–317. With regard to 4Q350, representing a record in Greek that lists quantities of cereals and being the re-used *verso* of 4Q460 9, Hannah Cotton and Erik Larson, "4Q460/4Q350 and Tampering with Qumran Texts in Antiquity?" in *Emanuel: Studies in Hebrew Bible, Septuagint, and Dead Sea Scrolls in Honor of Emanuel Tov*, ed. Shalom M. Paul, Robert A. Kraft, Lawrence H. Schiffman, and Weston W. Fields, Supplements to Vetus Testamentum 94 (Leiden: Brill, 2003), 113–25 argue that "the penning of an ephemeral list in Greek on the back of a sacred text in Hebrew points to non-Jewish occupants of the site" (122). But opistographs such as 4Q201 (4QEn^a ar) and its *verso* 4Q338 or the papyrus manuscripts 1Q70/1Q70bis and 4Q518/4Q519 indicate that manuscripts were reused and that different compositions were written on the *recto* and *verso* at an earlier stage, before the site of Qumran was destroyed and occupied by an auxiliary unit of the Roman army; see Mladen Popović, "Roman Book Destruction in Qumran Cave 4 and the Roman Destruction of Khirbet Qumran Revisited," in *Qumran und die Archäologie: Texte und Kontexte*, ed. Jörg Frey, Carsten Claußen, and Nadine Kessler, Wissenschaftliche Untersuchungen zum Neuen Testament 278 (Tübingen: Mohr Siebeck, 2011), 239–91, at 249; Matthew Richey, "The Use of Greek at Qumran: Manuscript and Epigraphic Evidence for a Marginalized Language," *Dead Sea Discoveries* 19 (2012): 177–97, at 184–86.

42 Joosten, "Hebrew, Aramaic, and Greek in the Qumran Scrolls," 369. For Greek loanwords in the Copper Scroll (3Q15), see Florentino García Martínez, "Greek Loanwords in the *Copper Scroll*," in *Qumranica Minora II: Thematic Studies on the Dead Sea Scrolls*, ed. Eibert J. C. Tigchelaar, Studies on the Texts of the Desert of Judah 64 (Leiden: Brill, 2007), 145–70. Greek epigraphic evidence points to the use of Greek in trade and economy, not only at Qumran but also elsewhere at Judaean Desert manuscript find sites; see, e.g., Richey, "The Use of Greek at Qumran." More generally on Greek in ancient Judaea, see Wise, *Language and Literacy*, 331–45.

43 Esther Eshel, "4QRebukes Reported by the Overseer," in *Qumran Cave 4.XXVI: Cryptic Texts and Miscellanea, Part 1*, ed. Philip Alexander et al., Discoveries in the Judaean Desert 36 (Oxford: Clarendon, 2000), 474–83.

44 On the spread of Greek names and what this may indicate about the spread of language proficiency in Greek, see Wise, *Language and Literacy*, 287.

45 Emanuel Tov, *Scribal Practices and Approaches*, 246–48.

characters were used as scribal markings in the margins of texts written in the square script.[46] In addition, in twenty-eight or twenty-nine scrolls, which are otherwise written in the square script, the four letters of the Hebrew name for God (the letters YHWH, also referred to as the Tetragrammaton) are not written in the square script – maybe out of respect or to prevent the name from accidentally being spoken when reading the text aloud – but in the paleo-Hebrew script.[47] Observing correlations within the corpus of Qumran scrolls, scholars understand "the use of palaeo-Hebrew characters for the divine name ... to be exclusive and characteristic for texts written according to the 'Qumran scribal practice' within the corpus."[48] A special link is suggested between the writing of the divine names in palaeo-Hebrew characters and the Qumran community.[49] Such a link may be suggestive, but there is evidence arguing against perceiving this practice of writing the divine name in palaeo-Hebrew as special to Qumran-specific manuscripts and a presumed Qumran community behind the scrolls. First, there are thirty-six manuscripts written in the so-called Qumran scribal practice that did not use this special system for writing the divine name.[50] Second, the phenomenon of writing the divine name in palaeo-Hebrew characters also occurs in Greek manuscripts, from elsewhere in the Judaean Desert (Naḥal Ḥever: 1/8ḤevXIIgr) and from Egypt (Oxyrhynchus: POxy 1007; POxy 3522).[51]

---

46 Tov, *Scribal Practices*, 206–8.
47 Tov, *Scribal Practices*, 238–46. See also Kristin De Troyer, "The Names of God, Their Pronunciation and Their Translation: A Digital Tour of Some of the Main Witnesses," *Lectio Difficilior* 2 (2005): http://www.lectio.unibe.ch/05_2/troyer_names_of_god.htm; Smelik, "*Ashurit* and Alphabet," 8–10.
48 Tigchelaar, "Assessing Emanuel Tov's 'Qumran Scribal Practice,'" 199–200.
49 See, e.g., recently Stephen Reed, "The Linguistic Diversity of the Texts Found at Qumran," in *The Scrolls from Qumran and the Concept of a Library*, ed. Sidnie White Crawford and Cecilia Wassén, Studies on the Texts of the Desert of Judah 116 (Leiden, 2016), 132–54, at 145.
50 Tov, *Scribal Practices*, 244.
51 The late first-century BCE Greek Minor Prophets Scroll from Naḥal Ḥever contains twenty-eight fully or partially preserved occurrences of the Tetragrammaton in palaeo-Hebrew; see Emanuel Tov, *The Greek Minor Prophets Scroll from Naḥal Ḥever (8ḤevXIIgr)*, Discoveries in the Judaean Desert 8 (Oxford: Clarendon, 1990). Apart from a number of finds dating to the Chalcolithic period, the refugee caves 5/6 and 8 of Naḥal Ḥever seem not to have been in use before the second century CE; see Popović, "Qumran as Scroll Storehouse," 561, 563; Yohanan Aharoni, "Expedition B – The Cave of Horror," *Israel Exploration Journal* 12 (1962): 186–99; Yigael Yadin, *The Finds from the Bar Kokhba Period in the Cave of the Letters*, Judean Desert Studies 1 (Jerusalem: Israel Exploration Society, 1963). The Greek Minor Prophets Scroll (1/8ḤevXIIgr) cannot, therefore, have made its way to Naḥal Ḥever before the Bar Kokhba revolt. This context indicates that Greek manuscripts with the Tetragrammaton in palaeo-Hebrew script circulated in a Jewish context until at least the first third of the second century CE. For the two examples from

Completely unexpected was the discovery of several script systems among the Dead Sea Scrolls, which were unknown until then. They have been conveniently named the "cryptic" scripts. Only Cryptic A has been deciphered thus far – the Cryptic B and C scripts, of which only a few fragments have been preserved, have not been deciphered.[52]

With regard to the use of multiple scripts in the Dead Sea Scrolls, we encounter a heterogeneity that resembles our findings with regard to multilingualism. The manuscript evidence attests to "pluriscriptural" identities and competencies. This is not to say that everyone had such "pluriscriptural" competencies, but the evidence especially of "mixed" or "multiscriptural" texts points further in the direction of an intellectual and scholarly identity for at least a number of people behind the scrolls from the caves near Qumran.

There are examples of Cryptic A texts that start with Hebrew in square script, such as 4Q298 (4QcryptA Words of the Maskil to All Sons of Dawn) and 4Q249 (4Qpap Crypt A Midrash Sefer Moshe).[53] And then there is a unique astrological and physiognomic manuscript, 4QZodiacal Physiognomy (4Q186). This exceptional manuscript from Qumran Cave 4 combines several scripts: square, palaeo-Hebrew, Greek, and Cryptic A. Moreover, it is written in reverse order: from left to right.[54]

The latter example of the "multiscriptural" text 4QZodiacal Physiognomy (4Q186) paves the way to see how elements of multilingualism and multiscripturalism in the Dead Sea Scrolls connect with knowledge transfer and strategies of sharing and hiding knowledge. This will allow us to understand how cultural encounters and language contact entangled the intellectuals behind the scrolls with other intellectual and textual communities in the ancient Mediterranean.

Oxyrhynchus, see, e.g., Emanuel Tov, *Textual Criticism of the Hebrew Bible, Second Revised Edition* (Minneapolis: Fortress and Assen: Van Gorcum: 2001), 220; De Troyer, "The Names of God."
52 Tov, *Scribal Practices*, 259–60. See now also Eshbal Ratzon and Jonathan Ben-Dov, "A Newly Reconstructed Calendrical Scroll from Qumran in Cryptic Script," *Journal of Biblical Literature* 136 (2017): 905–36.
53 Jonathan Ben-Dov and Daniel Stökl Ben Ezra, "4Q249 Midrash Moshe: A New Reading and Some Implications," *Dead Sea Discoveries* 21 (2014): 131–49.
54 For a more general discussion about the combination of several scripts and the writing in reverse order, see Popović, *Reading the Human Body*, 26, 227–30.

2.4 Knowledge Transfer

Scholars assume that the Cryptic A script was devised for the internal purposes of the community, but this is not evident.[55] This explanation often occurs in tandem with the notion of a group that is presumably isolated from its surroundings. As I explain in the introduction above, a selection of manuscripts has informed the scholarly construct of a Qumran sect or community as the sociological matrix for all manuscripts. Such a framework has influenced how other texts were contextualized within the scholarly narrative. Thus, the Cryptic A texts were categorized as "sectarian" and, for example, the Aramaic texts were not only seen as older than Hebrew non-biblical texts but also understood categorically as so-called "non-sectarian" texts. Thus, the Cryptic A script from Qumran has often been explained in terms of secrecy strategies, to hide learned knowledge from outsiders or insiders who were not fully initiated.

Specific strategies of sharing and hiding knowledge may indeed have been intended, but this perspective must now be divorced from the notion of an isolated and marginal community at the site of Khirbet Qumran. In order to better gauge the situation, two aspects need to be discussed: 1. The connections between learned knowledge in Graeco-Roman Judaea and scholarly knowledge elsewhere in the ancient Mediterranean and Near East, and the character of this connectivity; 2. The extent to which such scholarly literacy was spread and shared within Graeco-Roman Judaea.

A fascinating feature of the Dead Sea Scrolls is that they contain the oldest examples of scholarly writings within a Jewish context, such as astronomical, astrological, calendrical, and physiognomic texts.[56] In order to discuss the aspect of intercultural connectivity and its character, I will briefly consider one example.

---

55 Popović, *Reading the Human Body*, 10; Mladen Popović, "The Emergence of Aramaic and Hebrew Scholarly Texts: Transmission and Translation of Alien Wisdom," in *The Dead Sea Scrolls: Transmission of Traditions and Production of Texts*, ed. Sarianna Metso, Hindy Najman, and Eileen Schuller, Studies on the Texts of the Desert of Judah 92 (Leiden: Brill, 2010), 81–114, at 99–100, 105–6. The announced example of a cup discovered in 2009 during excavations in Jerusalem is as of yet unclear evidence for the occurrence of this script outside of a Qumran context; for images and an excerpt from the preliminary report, see http://popular-archaeology.com/issue/12012013/article/digging-into-first-century-jerusalem-s-rich-and-famous. See also Emanuel Tov, "Scribal Characteristics of the Qumran Scrolls," in *The Caves of Qumran: Proceedings of the International Conference, Lugano 2014*, ed. Marcello Fidanzio, Studies on the Texts of the Desert of Judah 118 (Leiden: Brill, 2017), 87–95, at 88.

56 For references and discussion, see Popović, "The Emergence of Aramaic and Hebrew Scholarly Texts"; Mladen Popović, "Networks of Scholars: The Transmission of Astronomical and Astrological Learning between Babylonians, Greeks and Jews," in *Ancient Jewish Sciences and the*

The Aramaic text 4QZodiology and Brontology (4Q318) consists of two parts. The first part (*selenodromion*) describes the synodic movement of the moon through the zodiac during twelve months of thirty days each, counting a 360-day year, as in Babylonian tradition. The second part (*brontologion*) has predictions for when it will thunder. This sort of text appears both in the Babylonian and Graeco-Roman astrological traditions.[57] The 360-day year scheme suggests a derivation from Babylonian tradition, but the zodiacal names in 4Q318 point to Hellenistic origins.[58] What direction of cultural influence does this text exemplify? Is the text to be taken as an example of the Aramaic language being a medium of transmission of Babylonian learning westward? But such a one-sided directionality makes it difficult to account for the Hellenistic elements in the text. The idea of cultural influence together with directionality seems to be the wrong approach for understanding the character of intercultural connectivity that the text of 4QZodiology and Brontology attests to. What we have here is a fascinating glimpse of a tradition that is not so easy to pinpoint for us. The text of 4QZodiology and Brontology (4Q318) indicates the existence of an Aramaic tradition of astrological and astronomical knowledge that circulated in the eastern Mediterranean between or within Babylonian and Hellenistic traditions, not unlike what we encounter in late antique and early medieval traditions in Syriac, Mandaic, Greek, Hebrew, and Arabic.

Scholars often assume that the Neo-Babylonian period was the moment when Judeans came into direct contact with learned knowledge from the Babylonian realm. The prophet Ezekiel from the Hebrew Bible is one example often put forward for such direct contact. However, in this case, and also with regard to astronomical, calendrical, astrological, and physiognomic learning from Qumran, the Neo-Babylonian period is an unlikely time-frame for that to have happened.[59] What we know of Babylonian culture at the time suggests that the elite was stricter

---

History of Knowledge in Second Temple Literature, ed. Jonathan Ben-Dov and Seth L. Sanders (New York: New York University Press, 2014), 151–91.
57 Popović, *Reading the Human Body*, 128.
58 See, e.g., Jonas C. Greenfield and Michael Sokoloff, "4QZodiology and Brontology ar," in *Qumran Cave 4.XXVI: Cryptic Texts and Miscellanea, Part 1*, ed. Philip Alexander et al., Discoveries in the Judaean Desert 36 (Oxford: Clarendon, 2000), 259–74; Reimund Leicht, *Astrologumena Judaica: Untersuchungen zur Geschichte der astrologischen Literatur der Juden*, Texts and Studies in Medieval and Early Modern Judaism 21 (Tübingen: Mohr Siebeck, 2006), 23–24; Helen R. Jacobus, *Zodiac Calendars in the Dead Sea Scrolls and Their Reception: Ancient Astronomy and Astrology in Early Judaism*, IJS Studies in Judaica 14 (Leiden: Brill, 2015).
59 Mladen Popović, "Ancient Jewish Cultural Encounters and a Case Study on Ezekiel," in *Jewish Cultural Encounters in the Ancient Mediterranean and Near Eastern World*, ed. Mladen Popović, Myles Schoonover, and Marijn Vandenberghe, Supplements to the Journal for the Study of Judaism 178 (Leiden: Brill, 2017), 1–12.

in maintaining their boundaries with regard to cuneiform culture and their learned knowledge, not sharing it with those belonging to non-Babylonian elites.[60]

In the Late Babylonian period, however, we have clear evidence – literary, documentary, and epigraphical – of the transfer of learned knowledge from the Babylonian to the Hellenistic realm. We also have possible evidence of Aramaic scribes (*sepiru*) involved in the production of Babylonian scientific texts on scrolls,[61] showing that what seemed an impermeable boundary in the Neo-Babylonian period between different kinds of scribes in relation to different kinds of textual and intellectual production was not so anymore in the Late Babylonian period.

With regard to the most likely time-frame for a type of connectivity to exist that enabled direct or indirect knowledge transfer from the Babylonian realm westward it is interesting to consider for a moment the ongoing creation of literary traditions about the Neo-Babylonian king Nabonidus. Caroline Waerzeggers argues that cuneiform historical literature from the late Persian and Hellenistic periods invites us for the Aramaic Nabonidus material at Qumran "to rethink models that assume centuries of transmission through texts, memories, or both. Instead, these materials offer the possibility of considering a more collateral, synchronic development—one that engaged literary communities across regions." The lively and productive debate about Nabonidus that Babylonian scholars in the Hellenistic period were engaged in may have provided a general cultural context, Waerzeggers suggests, in which Babylonian-Jewish interactions that are behind such literary texts as Daniel 4 or the Prayer of Nabonidus may well have occurred, rather in a third or second century BCE context than three centuries earlier.[62]

Thus, a time-frame in the Hellenistic period would be plausible for the existence of an intellectual connectivity between Graeco-Roman Judaea, Hellenistic Babylonia, and the eastern Mediterranean. That cultural context may have provided the circumstances most conducive for the exchange of scholarly literacy exemplified by the astronomical, astrological, calendrical, and physiognomic texts from Qumran. These texts attest to direct or indirect contact between intel-

---

[60] For references and discussion, see Popović, "Networks of Scholars," 169–74. See also Eleanor Robson's contribution in this volume.
[61] For six references to such *magallatu* ("scrolls"), see Seth L. Sanders, *From Adapa to Enoch: Scribal Culture and Religious Vision in Judea and Babylon*, Texts and Studies in Ancient Judaism 167 (Tübingen: Mohr Siebeck, 2017), 188–94.
[62] Caroline Waerzeggers, "The *Prayer of Nabonidus* in the Light of Hellenistic Babylonian Literature," in *Jewish Cultural Encounters in the Ancient Mediterranean and Near Eastern World*, ed. Mladen Popović, Myles Schoonover, and Marijn Vandenberghe, Supplements to the Journal for the Study of Judaism 178 (Leiden: Brill, 2017), 64–75, the quote is from 65.

lectuals and scholars in Graeco-Roman Judaea and intellectuals and scholars from elsewhere in the ancient Mediterranean.

If indeed intellectuals and scholars in Graeco-Roman Judaea were connected with broader, international developments of scholarly learning, possibly via an Aramaic learned tradition that circulated in the eastern Mediterranean, to what extent was such scholarly learning accessible within Graeco-Roman Judaea? Invoking Pierre Bourdieu's concept of cultural capital, we can appreciate the scholarly literacy, multilingualism,[63] and multiscripturalism evinced by a number of the Qumran texts to have been perceived as prized pieces of knowledge signalling and confirming the status of those having access to and possessing it.[64] On one level, that of international ancient Mediterranean scholarly literacy, the evidence does not invite us to construct a dichotomy between Aramaic and Hebrew, Babylonian, Hellenistic, and Judaean. On another level, differentiating within Graeco-Roman Judaea between different types of literacy, explicit and implicit boundaries will have controlled the dissemination of and access to this scholarly learning. Even if having had the chance to read one, Aramaic literary texts will have been difficult to grasp for many.[65] A Cryptic A text would probably proof difficult even for those who had attained a fluent level of literary literacy, and a mixed and reversely written text such as 4QZodiacal Physiognomy (4Q186) even more so.

In Babylonia, secrecy formulae appear as important topoi in the colophons of scholarly texts. They signal a form of boundary maintenance with regard to literary and scholarly texts, limiting the accessibility and mobility of scholarly knowledge. Instead of interpreting these secrecy formulae as signals for a single, abstract body of "secret knowledge," Kathryn Stevens has argued to consider these formulae as a type of protective mechanism chosen by Babylonian scholars to protect texts closely associated with their scholarly expertise and their personal professional identity.[66] For Graeco-Roman Judaea, we have no evidence for such colophons,[67] but the mechanism of intellectual protection because of a

---

[63] One may think of the relationship between the Hebrew astrological and physiognomic text 4Q186 and the Aramaic physiognomic text 4Q561; see Popović, "The Emergence of Aramaic and Hebrew Scholarly Texts," 103–6.
[64] Mladen Popović, "Physiognomic Knowledge in Qumran and Babylonia: Form, Interdisciplinarity, and Secrecy," *Dead Sea Discoveries* 13 (2006): 150–76, at 166–76; Popović, *Reading the Human Body*, 81–83, 100, 227–31.
[65] Wise, *Language and Literacy*, 317–31.
[66] Kathryn Stevens, "Secrets in the Library: Protected Knowledge and Professional Identity in Late Babylonian Uruk," *Iraq* 75 (2013): 211–53. See also Eleanor Robson's contribution in this volume.
[67] Mladen Popović, "Pseudepigraphy and a Scribal Sense of the Past in the Ancient Mediterranean: A Copy of the Book of the Words of the Vision of Amram," in *Is There a Text in This Cave?*

correlation between scholarly expertise and personal professional identity may have been in operation as well. In Graeco-Roman Judaea, specific strategies for sharing or hiding that knowledge were in operation by means of multilingualism, multiscripturalism, and scholarly literacy.

The concrete manifestations of knowledge transfer and connectivity in which intellectuals and scholars in Graeco-Roman Judaea took part – on the ancient Mediterranean level and on the level of Graeco-Roman Judaea – argue against characterising the people behind the Dead Sea Scrolls as an isolated and marginal community. Rather, the scrolls provide us with unique access to a textual and intellectual community that can function as a lens through which we can observe a rich intellectual, multilingual, and multiscriptural world that not only connected but embodied elements from Hellenistic, Babylonian, and Judaean cultures of writing and learning.

References

Martin G. Abegg, "Linguistic Profiles of the Isaiah Scrolls," in *Qumran Cave 1.II, The Isaiah Scrolls, Part 2: Introductions, Commentary, and Textual Variants*, Eugene Ulrich and Peter W. Flint, Discoveries in the Judaean Desert 32 (Oxford: Clarendon, 2010), 25–41.

James N. Adams, *Bilingualism and the Latin Language* (Cambridge: Cambridge University Press, 2003).

James N. Adams, Mark Janse, and Simon Swain, eds., *Bilingualism in Ancient Society: Language Contact and the Written Text* (Oxford: Oxford University Press, 2002).

James N. Adams and Simon Swain, "Introduction," in *Bilingualism in Ancient Society: Language Contact and the Written Text*, ed. James N. Adams, Mark Janse, and Simon Swain (Oxford: Oxford University Press, 2002), 1–20.

Yohanan Aharoni, "Expedition B – The Cave of Horror," *Israel Exploration Journal* 12 (1962): 186–99.

Kormi Anipa, "The Use of Literary Sources in Historical Sociolinguistic Research," in *The Handbook of Historical Sociolinguistics*, ed. Juan Manuel Hernández-Campoy and Juan Camilo Conde-Silvestre (Oxford: Wiley-Blackwell, 2012), 170–90.

Albert I. Baumgarten, *The Flourishing of Jewish Sects in the Maccabean Era: An Interpretation*, Supplements to the Journal for the Study of Judaism 55 (Leiden: Brill, 1997).

Robert Bayley, Richard Cameron, and Ceil Lucas, eds., *The Oxford Handbook of Sociolinguistics* (Oxford: Oxford University Press, 2013).

Jonathan Ben-Dov, *Head of All Years: Astronomy and Calendars at Qumran in their Ancient Context* (Leiden: Brill, 2008).

Studies in the Textuality of the Dead Sea Scrolls in Honour of George J. Brooke, ed. Ariel Feldman, Maria Cioată, and Charlotte Hempel, Studies on the Texts of the Desert of Judah 119 (Leiden: Brill, 2017), 308–18.

Jonathan Ben-Dov and Seth L. Sanders, eds., *Ancient Jewish Sciences and the History of Knowledge in Second Temple Literature* (New York: New York University Press, 2014).

Jonathan Ben-Dov and Daniel Stökl Ben Ezra, "4Q249 Midrash Moshe: A New Reading and Some Implications," *Dead Sea Discoveries* 21 (2014): 131–49.

James Clackson, "Language Maintenance and Language Shift in the Mediterranean World during the Roman Empire," in *Multilingualism in the Graeco-Roman Worlds*, ed. Alex Mullen and Patrick James (Cambridge: Cambridge University Press, 2012), 36–57.

Hannah Cotton and Erik Larson, "4Q460/4Q350 and Tampering with Qumran Texts in Antiquity?" in *Emanuel: Studies in Hebrew Bible, Septuagint, and Dead Sea Scrolls in Honor of Emanuel Tov*, ed. Shalom M. Paul, Robert A. Kraft, Lawrence H. Schiffman, and Weston W. Fields, Supplements to Vetus Testamentum 94 (Leiden: Brill, 2003), 113–25.

Hannah Cotton, Robert G. Hoyland, Jonathan J. Price, and David J. Wasserstein, eds., *From Hellenism to Islam: Cultural and Linguistic Change in the Roman Near East* (Cambridge: Cambridge University Press, 2009).

Hannah Cotton and Ada Yardeni, *Aramaic, Hebrew and Greek Documentary Texts from Naḥal Ḥever and Other Sites, with an Appendix Containing Alleged Qumran Texts*, Discoveries in the Judaean Desert 27 (Oxford: Clarendon, 2010).

Florian Coulmas, *Writing and Society: An Introduction* (Cambridge: Cambridge University Press, 2012).

Esther Eshel, "4QRebukes Reported by the Overseer," in *Qumran Cave 4.XXVI: Cryptic Texts and* Miscellanea, *Part 1*, ed. Philip Alexander et al., Discoveries in the Judaean Desert 36 (Oxford: Clarendon, 2000), 474–83.

Steven D. Fraade, "Language Mix and Multilingualism in Ancient Palestine: Literary and Inscriptional Evidence," *Jewish Studies* 48 (2012): 1–40.

Florentino García Martínez, "Greek Loanwords in the *Copper Scroll*," in *Qumranica Minora II: Thematic Studies on the Dead Sea Scrolls*, ed. Eibert J. C. Tigchelaar, Studies on the Texts of the Desert of Judah 64 (Leiden: Brill, 2007), 145–70.

Jonas C. Greenfield and Michael Sokoloff, "4QZodiology and Brontology ar," in *Qumran Cave 4.XXVI: Cryptic Texts and* Miscellanea*, Part 1*, ed. Philip Alexander et al., Discoveries in the Judaean Desert 36 (Oxford: Clarendon, 2000), 259–74.

Pieter B. Hartog, *Pesher and Hypomnema: A Comparison of Two Commentary Collections from the Hellenistic-Roman Period*, Studies on the Texts of the Desert of Judah 121 (Leiden: Brill, 2017).

Charlotte Hempel, *The Qumran Rule Texts in Context: Collected Studies*, Texts and Studies in Ancient Judaism 154 (Tübingen: Mohr Siebeck, 2013).

Martin Hengel, "Qumrān und der Hellenismus," in *Qumrân: Sa piété, sa théologie et son milieu*, ed. Matthias Delcor, Bibliotheca Ephemeridum Theologicarum Lovaniensium 46 (Paris: Duculot and Leuven: Leuven University Press, 1978), 333–72.

Catherine Hezser, *Jewish Literacy in Roman Palestine*, Texts and Studies in Ancient Judaism 81 (Tübingen: Mohr Siebeck, 2001).

Helen R. Jacobus, *Zodiac Calendars in the Dead Sea Scrolls and Their Reception: Ancient Astronomy and Astrology in Early Judaism*, IJS Studies in Judaica 14 (Leiden: Brill, 2015).

Jan Joosten, "Hebrew, Aramaic, and Greek in the Qumran Scrolls," in *The Oxford Handbook of the Dead Sea Scrolls*, ed. Timothy H. Lim and John J. Collins (Oxford: Oxford University Press, 2010), 351–74.

Edward Y. Kutscher, *The Language and Linguistic Background of the Isaiah Scroll (1QIsa^a)*, Studies on the Texts of the Desert of Judah 6 (Leiden: Brill, 1974).

Reimund Leicht, *Astrologumena Judaica: Untersuchungen zur Geschichte der astrologischen Literatur der Juden*, Texts and Studies in Medieval and Early Modern Judaism 21 (Tübingen: Mohr Siebeck, 2006).

Timothy H. Lim, "The Qumran Scrolls, Multilingualism, and Biblical Interpretation," in *Religion in the Dead Sea Scrolls*, ed. John J. Collins and Robert A. Kugler (Grand Rapids: Michigan, 2000), 57–73.

Fergus Millar, "Introduction: Documentary Evidence, Social Realities and the History of Language," in *From Hellenism to Islam: Cultural and Linguistic Change in the Roman Near East*, ed. Hannah M. Cotton, Robert G. Hoyland, Jonathan J. Price, and David J. Wasserstein (Cambridge: Cambridge University Press, 2009), 1–12.

Alex Mullen, *Southern Gaul and the Mediterranean: Multilingualism and Multiple Identities in the Iron Age and Roman Periods* (Cambridge: Cambridge University Press, 2013).

Alex Mullen and Patrick James, eds., *Multilingualism in the Graeco-Roman Worlds* (Cambridge: Cambridge University Press, 2012).

Joseph Naveh, *Early History of the Alphabet* (Jerusalem: Magnes and Leiden: Brill, 1982).

G. Wilhelm Nebe, "Das Sprachvermögen des Mebaqqer in *Damaskusschrift* XIV, 10," *Revue de Qumrân* 16/62 (1993): 289–91.

Jacobine G. Oudshoorn, *The Relationship between Roman and Local Law in the Babatha and Salome Komaise Archives: General Analysis and Three Case Studies on the Law of Succession, Guardianship and Marriage*, Studies on the Texts of the Desert of Judah 69 (Leiden: Brill, 2007).

Sari Pietikäinen and Helen Kelly-Holmes, eds., *Multilingualism and the Periphery* (Oxford: Oxford University Press, 2013).

Sari Pietikäinen and Helen Kelly-Holmes, "Multilingualism and the Periphery," in *Multilingualism and the Periphery*, ed. Sari Pietikäinen and Helen Kelly-Holmes (Oxford: Oxford University Press, 2013), 1–16.

Mladen Popović, "Physiognomic Knowledge in Qumran and Babylonia: Form, Interdisciplinarity, and Secrecy," *Dead Sea Discoveries* 13 (2006): 150–76.

Mladen Popović, *Reading the Human Body: Physiognomics and Astrology in the Dead Sea Scrolls and Hellenistic-Early Roman Period Judaism*, Studies on the Texts of the Desert of Judah 67 (Leiden: Brill, 2007).

Mladen Popović, "The Emergence of Aramaic and Hebrew Scholarly Texts: Transmission and Translation of Alien Wisdom," in *The Dead Sea Scrolls: Transmission of Traditions and Production of Texts*, ed. Sarianna Metso, Hindy Najman, and Eileen Schuller, Studies on the Texts of the Desert of Judah 92 (Leiden: Brill, 2010), 81–114.

Mladen Popović, "Roman Book Destruction in Qumran Cave 4 and the Roman Destruction of Khirbet Qumran Revisited," in *Qumran und die Archäologie: Texte und Kontexte*, ed. Jörg Frey, Carsten Claußen, and Nadine Kessler, Wissenschaftliche Untersuchungen zum Neuen Testament 278 (Tübingen: Mohr Siebeck, 2011), 239–91.

Mladen Popović, "Qumran as Scroll Storehouse in Times of Crisis? A Comparative Perspective on Judaean Desert Manuscript Collections," *Journal for the Study of Judaism* 43 (2012): 551–94.

Mladen Popović, "Networks of Scholars: The Transmission of Astronomical and Astrological Learning between Babylonians, Greeks and Jews," in *Ancient Jewish Sciences and the History of Knowledge in Second Temple Literature*, ed. Jonathan Ben-Dov and Seth L. Sanders (New York: New York University Press, 2014), 151–91.

Mladen Popović, "Scribal Culture of the Hebrew Bible and the Burden of the Canon: Human Agency and Textual Production and Consumption in Ancient Judaism," in *Jeremiah's*

Scriptures: Production, Reception, Interaction, and Transformation, ed. Hindy Najman and Konrad Schmid, Supplements to the Journal for the Study of Judaism 173 (Leiden: Brill, 2016), 253–58.

Mladen Popović, "Ancient Jewish Cultural Encounters and a Case Study on Ezekiel," in *Jewish Cultural Encounters in the Ancient Mediterranean and Near Eastern World*, ed. Mladen Popović, Myles Schoonover, and Marijn Vandenberghe, Supplements to the Journal for the Study of Judaism 178 (Leiden: Brill, 2017), 1–12.

Mladen Popović, "Pseudepigraphy and a Scribal Sense of the Past in the Ancient Mediterranean: A Copy of the Book of the Words of the Vision of Amram," in *Is There a Text in This Cave? Studies in the Textuality of the Dead Sea Scrolls in Honour of George J. Brooke*, ed. Ariel Feldman, Maria Cioată, and Charlotte Hempel, Studies on the Texts of the Desert of Judah 119 (Leiden: Brill, 2017), 308–18.

Mladen Popović, Myles Schoonover, and Marijn Vandenberghe, eds., *Jewish Cultural Encounters in the Ancient Mediterranean and Near Eastern World*, Supplements to the Journal for the Study of Judaism 178 (Leiden: Brill, 2017).

Eshbal Ratzon and Jonathan Ben-Dov, "A Newly Reconstructed Calendrical Scroll from Qumran in Cryptic Script," *Journal of Biblical Literature* 136 (2017): 905–36.

Stephen Reed, "The Linguistic Diversity of the Texts Found at Qumran," in *The Scrolls from Qumran and the Concept of a Library*, ed. Sidnie White Crawford and Cecilia Wassén, Studies on the Texts of the Desert of Judah 116 (Leiden, 2016), 132–54.

Gary A. Rendsburg, "Qumran Hebrew (With a Trial Cut [1QS])," in *The Dead Sea Scrolls at 60: Scholarly Contributions of New York University Faculty and Alumni*, ed. Lawrence H. Schiffman and Shani Tzoref, Studies on the Texts of the Desert of Judah 89 (Leiden: Brill, 2010), 217–46.

Matthew Richey, "The Use of Greek at Qumran: Manuscript and Epigraphic Evidence for a Marginalized Language," *Dead Sea Discoveries* 19 (2012): 177–97.

Hanna Rutkowska and Paul Rössler, "Orthographic Variables," in *The Handbook of Historical Sociolinguistics*, ed. Juan Manuel Hernández-Campoy and Juan Camilo Conde-Silvestre (Oxford: Wiley-Blackwell, 2012), 213–36.

Seth L. Sanders, *From Adapa to Enoch: Scribal Culture and Religious Vision in Judea and Babylon*, Texts and Studies in Ancient Judaism 167 (Tübingen: Mohr Siebeck, 2017).

Herbert Schendl, "Multilingualism, Code-switching, and Language Contact in Historical Sociolinguistics," in *The Handbook of Historical Sociolinguistics*, ed. Juan Manuel Hernández-Campoy and Juan Camilo Conde-Silvestre (Oxford: Wiley-Blackwell, 2012), 520–33.

William H. Schniedewind, "Qumran Hebrew as an Antilanguage," *Journal of Biblical Literature* 118 (1999): 235–52.

William H. Schniedewind, "Linguistic Ideology in Qumran Hebrew," in *Diggers at the Well: Proceedings of a Third International Symposium on the Hebrew of the Dead Sea Scrolls and Ben Sira*, ed. Takamitsu Muraoka and John F. Elwolde, Studies on the Texts of the Desert of Judah 36 (Leiden: Brill, 2000), 245–55.

William H. Schniedewind, *A Social History of Hebrew: Its Origins Through the Rabbinic Period* (New Haven: Yale University Press, 2013).

Willem Smelik, "Code-switching: The Public Reading of the Bible in Hebrew, Aramaic and Greek," in *Was ist ein Text? Alttestamentliche, ägyptologische und altorientalische Perspektiven*, ed. Ludwig Morenz and Stefan Schorch (Berlin: De Gruyter, 2007), 123–47.

Willem Smelik, "The Languages of Roman Palestine," in *The Oxford Handbook of Jewish Daily Life in Roman Palestine*, ed. Catherine Hezser (Oxford: Oxford University Press, 2010), 122–41.

Willem Smelik, "Holy Tongue," in *Rabbis, Language and Translation in Late Antiquity* (Cambridge: Cambridge University Press, 2013), 42–99.

Willem Smelik, "*Ashurit* and Alphabet," in *Rabbis, Language and Translation in Late Antiquity* (Cambridge: Cambridge University Press, 2013), 271–322.

Bernard Spolsky, "Jewish Multilingualism in the First Century: An Essay in Historical Sociolinguistics," in *Readings in the Sociology of Jewish Languages*, ed. Joshua A. Fishman (Leiden: Brill, 1985), 34–50.

Kathryn Stevens, "Secrets in the Library: Protected Knowledge and Professional Identity in Late Babylonian Uruk," *Iraq* 75 (2013): 211–53.

Brian Stock, *The Implications of Literacy: Written Language and Models of Interpretation in the Eleventh and Twelfth Centuries* (Princeton: Princeton University Press, 1983).

Brian Stock, *Listening for the Text: On the Uses of the Past* (Baltimore: Johns Hopkins University Press, 1990).

Dorothy J. Thompson, "The Multilingual Environment of Persian and Ptolemaic Egypt: Egyptian, Aramaic, and Greek Documentation," in *The Oxford Handbook of Papyrology*, ed. Roger S. Bagnall (Oxford: Oxford University Press, 2009), 395–417.

Eibert J. C. Tigchelaar, "Assessing Emanuel Tov's 'Qumran Scribal Practice,'" in *The Dead Sea Scrolls: Transmission of Traditions and Production of Texts*, ed. Sarianna Metso, Hindy Najman, and Eileen Schuller, Studies on the Texts of the Desert of Judah 92 (Leiden: Brill, 2010), 173–205.

Eibert J. C. Tigchelaar, "The Material Variance of the Dead Sea Scrolls: On Texts and Artefacts," *HTS Teologiese Studies/Theological Studies* 72/4 (2016): 1–6.

Emanuel Tov, *The Greek Minor Prophets Scroll from Naḥal Ḥever (8ḤevXIIgr)*, Discoveries in the Judaean Desert 8 (Oxford: Clarendon, 1990).

Emanuel Tov, *Textual Criticism of the Hebrew Bible, Second Revised Edition* (Minneapolis: Fortress and Assen: Van Gorcum: 2001).

Emanuel Tov, *Scribal Practices and Approaches Reflected in the Texts Found in the Judean Desert*, Studies on the Texts of the Desert of Judah 54 (Leiden: Brill, 2004).

Emanuel Tov, "Scribal Practices and Approaches Revisited," *Hebrew Bible and Ancient Israel* 3 (2014): 355–67.

Emanuel Tov, "Scribal Characteristics of the Qumran Scrolls," in *The Caves of Qumran: Proceedings of the International Conference, Lugano 2014*, ed. Marcello Fidanzio, Studies on the Texts of the Desert of Judah 118 (Leiden: Brill, 2017), 87–95.

Kristin De Troyer, "The Names of God, Their Pronunciation and Their Translation: A Digital Tour of Some of the Main Witnesses," *Lectio Difficilior* 2 (2005): http://www.lectio.unibe.ch/05_2/troyer_names_of_god.htm.

Marcus K. M. Tso, *Ethics in the Qumran Community: An Interdisciplinary Investigation*, Wissenschaftliche Untersuchungen zum Neuen Testament 292 (Tübingen: Mohr Siebeck, 2010).

David Vanderhooft, "'el-mĕdînâ ûmĕdînâ kiktābāh: Scribes and Scripts in Yehud and in Achaemenid Transeuphratene," in *Judah and the Judeans in the Achaemenid Period: Negotiating Identity in an International Context*, ed. Oded Lipschits, Gary N. Knoppers, and Manfred Oeming (Winona Lakes: Eisenbrauns, 2011), 529–44.

Caroline Waerzeggers, "The *Prayer of Nabonidus* in the Light of Hellenistic Babylonian Literature," in *Jewish Cultural Encounters in the Ancient Mediterranean and Near Eastern World*, ed. Mladen Popović, Myles Schoonover, and Marijn Vandenberghe, Supplements to the Journal for the Study of Judaism 178 (Leiden: Brill, 2017), 64–75.

Steven Weitzman, "Why Did the Qumran Community Write in Hebrew?" *Journal of the American Oriental Society* 119 (1999): 35–45.

Michael O. Wise, *Language and Literacy in Roman Judaea: A Study of the Bar Kokhba Documents* (New Haven: Yale University Press, 2015).

Yigael Yadin, *The Finds from the Bar Kokhba Period in the Cave of the Letters*, Judean Desert Studies 1 (Jerusalem: Israel Exploration Society, 1963).

Jacques van Ruiten
3 Sharing and Hiding Religious Knowledge in the Book of Jubilees

3.1 Introduction

In this contribution, I would like to focus on the book of Jubilees, written some time in the second century BCE.[1] It is a rewriting and interpretation of the books of Genesis and part of Exodus.[2] I discuss the topic of the transfer of knowledge within the narrative of this book. Given the fact that the book expresses, in its narrative, a strong cultural and religious resistance against other people, the question of the transmission of knowledge from and to other people is very

---

[1] Most influential has been the view of James C. VanderKam, who argued that the time of composition was mid-second century B.C.E. External evidence suggests that the oldest Qumran fragment (4Q216) may date to 125–100 BCE. Moreover, Jubilees is mentioned by name in the Damascus Document (CD 16:2–4), the earliest copy of which can be dated to 100–50 BCE. The internal evidence that is pointed to is the use of all existing parts of 1 Enoch, which would point to the fact that the book of Jubilees had been composed after the Book of Dreams, which was written in 164 or 163 BCE. Since there is no evidence in the book in favour of withdrawal from Jewish society, it may have been written prior to the foundation of the community of Qumran (150–140 BCE.). See James C. VanderKam, "Recent Scholarship on the Book of *Jubilees*," *Currenst in Research: Biblical Studies* 6 (2008): 405–31. One can question the sustainability of these arguments. Since only a very small percentage of the whole work is represented in the scrolls, it is difficult to draw firm conclusions as to the exact form and size of the book in the Second Temple period. It is doubtful whether any of the surviving fragments could have contained the whole book. Moreover, in my opinion, the many parallels between Jubilees and 1 Enoch do not point to the literary dependence of Jubilees on the text of the Book of Dreams. See, e.g., Jacques T. A. G. M. van Ruiten, "A Literary Dependency of *Jubilees* on *1 Enoch*: A Reassessment of a Thesis of J.C. VanderKam, *Henoch* 26 (2004): 205–9; Michael Knibb, "Which Parts of *1 Enoch* Were Known to *Jubilees*? A Note on the Interpretation of *Jub.* 4.16–25," in *Reading from Right to Left: Essays on the Hebrew Bible in Honour of David J.A. Clines*, ed. J. Cheryl Exum and Hugh G. M. Williamson, Journal for the Study of the Old Testament Supplement Series 373 (London: Sheffield Academic Press, 2003), 254–62. Therefore, other possibilities cannot be ruled out. Some scholars opt for a pre-Hasmonean date, since the book does not mention the persecution and decrees of Antiochus IV. See, e.g., George W. E. Nickelsburg, *Jewish Literature between the Bible and the Mishnah: A Historical and Literary Introduction*, 2nd ed. (Minneapolis: Fortress, 2005), 73–74. A few others argue for a date late in the second century because of the close similarities between Jubilees and the Qumran texts. See, e.g., Cana Werman, "The Book of *Jubilees* and the Qumran Community: The Relationship between the Two," *Megillot* 2 (2004): 37–55 [Hebrew]; Martha Himmelfarb, *A Kingdom of Priests: Ancestry and Merit in Ancient Judaism* (Philadelphia: University of Pennsylvania Press, 2006), 80–83.
[2] Cf. footnote 4.

relevant. I will show that there is a strong opposition against the transmission of knowledge from foreign people. There is no opposition to the transfer of the knowledge of Israel to other people, although this knowledge is meant in the first place for the own people. It is transmitted in a gradual and continuous process from the creation onwards. Originally, it is taught by the angels, who learned this from the heavenly tablets. Angels transmit this supernatural knowledge to distinguished patriarchs, who in turn transmit it to their sons and, ultimately, to all of the children of Israel.

3.2 Resistance against Foreign Knowledge

The question of the transmission of knowledge is of special interest, since the author of the book of Jubilees creates a strong dichotomy between the nations, and the nation par excellence, Israel. The author is erecting sharp boundaries between Israel and the other people, between insiders and outsiders. Israel is summoned to separate from the nations. The people of Israel should stay far away from them, and from their customs and their practices. There are, of course, earlier Jewish works who establish a culpability of the gentiles, and therefore a justification of divine punishment of the nations, as for example in Liber Antiquitatum Biblicarum, the Apocalypse of Abraham and the Testament of Moses. But the book of Jubilees occupies by far the most extreme position at the negative side of the spectrum. For the author of this work, the only way to please God is to keep the Torah, the ancestral laws, in its totality.[3]

This anti-gentile bias plays a role everywhere in the book, especially in material that is added to the rewriting of Gen 1–Exod 19.[4] Most texts that speak about

---

[3] Terence L. Donaldson, *Judaism and the Gentiles: Jewish Patterns of Universalism (to 135 CE)* (Waco: Baylor University Press, 2007). See also Martin Goodman, George H. van Kooten, and Jacques T. A. G. M. van Ruiten, eds., *Abraham, the Nations, and the Hagarites: Jewish, Christian, and Islamic Perspectives on Kinship with Abraham*, Themes in Biblical Narrative 13 (Leiden: Brill, 2010).

[4] Even if we accept the view that it is not appropriate to characterize *Jubilees* as an example of rewritten scripture – because the final form of scripture (as canonical closure of the Pentateuch) was not yet fixed at the time of Jubilees – it is still apparent that Jubilees presupposes most of the material that can be found in the text of Gen 1 to Exod 19 (cf. also Exod 24). See Hindy Najman and Eibert Tigchelaar, "Unity after Fragmentation," *Revue de Qumrân* 104/26 (2014): 495–500. In fact, they suggest that scholars who consider Jubilees as an example of rewritten scriptures work with three assumptions: (1) there is already a more or less stable biblical narrative; (2) much of the rewritten bible is driven by exegetical questions; (3) a given work of rewritten bible stems from the solutions to those problems proposed by a single writer (p. 496). They would like to

the erection of the boundaries between Israel and its environment can be found in the so-called farewell speeches, in the testaments of Noah, Abraham, Rebekah and Isaac, uttered just before the patriarchs die. In these speeches, it are the sons of the patriarchs who are addressed, and it is especially Jacob who plays an important role.

Jubilees creates strong borders between the people of Israel and the other groups. This means the exclusion of other groups from their own group. The reason that is given for this strong exclusion is that the habits and practices of the other groups are not the right ones for Israel. On the contrary contact with idolatrous non-Israelites is a threat to the religious belief of Israel; therefore, many Jews opted for a limitation of social interaction with gentiles. The separation from the nations prevents Israel from imitating their "actions" and "ways" and "worship." In Abraham's farewell speech to his grandson, Jacob, just before his death (Jub. 22:16–22), Abraham reinforces the requirement to separate from the nations, which is realized by abstinence from common meals, by not concluding agreements with them, and by keeping from intermarriage. This separation is ultimately related to the prohibition against idolatry, which corresponds to the worship of the God of Israel alone. With the covenant, this unique God was considered to have made Israel His partner from creation onwards. In Jubilees, there is one unique and eternal covenant between God and His chosen people, and this establishes Israel as different from all other peoples.[5]

Right from the beginning, we might say that there is a strong cultural and religious resistance against other people, and there is a strong opposition against the transmission of the knowledge from foreign peoples to Israel. In fact, a hiding of this knowledge for the people of Israel is required.

What we do not see is the hiding of the religious knowledge of Israel from other people. It is true that the revelation of knowledge is directed towards Israel,

---

suspend the first and third assumptions, in the sense that there was no canonical closure of the Pentateuch at that time. Therefore, they would like to see Jubilees as part of an ongoing tradition (p. 497). For a comparison of Gen 1–11 and Jub. 2–10, see Jacques T. A. G. M. van Ruiten, *Primaeval History Interpreted: The Rewriting of Genesis 1–11 in the Book of* Jubilees, Supplements to the Journal for the Study of Judaism 66 (Leiden: Brill, 2000); for a comparison of Gen 12–25 and Jub. 11–23, see idem, *Abraham in the Book of Jubilees: The Rewriting of Genesis 11:26–25:10 in the Book of Jubilees 11:14–23:8*, Supplements to the Journal for the Study of Judaism 161 (Leiden: Brill, 2012). For a comparison of the Jacob story in Genesis and Jubilees, see John C. Endres, *Biblical Interpretation in the Book of* Jubilees, Catholic Biblical Quarterly Monograph Series 18 (Washington: Catholic Biblical Association of America, 1987). For the Exodus material in *Jubilees*, see Betsy Halpern-Amaru, *The Perspective from Mt. Sinai: The Book of* Jubilees *and Exodus*, Journal of Ancient Judaism Supplements 21 (Göttingen: Vandenhoeck & Ruprecht, 2015).

5 See Van Ruiten, *Abraham in the Book of* Jubilees, 309–18.

and not so much to the nations. On several places, it is said that Moses commands and makes known to Israel: "Now you command the Israelites" (Jub. 2:26; 6:13, 20, 32; 15:28); "Inform and tell the Israelites" (Jub. 2:29), Now you, Moses, order the Israelites and testify to them" (Jub. 30:11; cf. 33:13, 18; 49:22). This mostly halakic knowledge is meant for Israel, and therefore implicitly not for the nations, but it is nowhere said that this knowledge should be hided from other people. It is even explicitly said that certain knowledge is meant for all people. In Jub. 4, Enoch, as the first of mankind, wrote down in a book the signs of the sky so that mankind would know the seasons of the years according to the fixed patterns of each of their month. He testified to mankind in the generations of the earth. In Jub. 1 it can be read that at the end of time, when the sanctuary is rebuilt in the midst of Israel, "the Lord will appear in the sight of all. And everyone will know that I am the God of Israel and the father of all the children of Jacob and king upon Mount Zion forever and ever" (Jub. 1:27–28).

3.3 Transmission of Knowledge within the Family

In Jubilees, transmission of knowledge does not have an esoteric allure. There is not an implicit nor an explicit address to a reduced circle of "initiated." Knowledge is not intended for a reduced group of followers, but for the whole of Israel, even for all humanity. In the narrative of Jubilees, one can nevertheless detect a certain opinion about the transmission of knowledge. There is a kind of transmission concerning religious knowledge, in which not only God and the angels, but also father and son are involved. The people of Israel gain knowledge in a gradual process and there is a strong emphasis on the continuity of teaching and learning from creation onwards. This transmission is related to a system of writing. The text speaks about fathers teaching their sons. At the same time, right from the beginning, one discovers figures from the heavenly world (God/angel) who teach Moses and the patriarchs about heavenly knowledge.[6]

In Jub. 4:17–26, Enoch is described as "the first of mankind who were born on the earth who learned (the art of) writing, instruction, and wisdom" (Jub. 4:17).[7]

---

[6] In a slightly different form, some of the results of this research are incorporated in my "Interpreting Torah: Jubilees at the Interface between Education and Religion," in *Schriftauslegung im Spannungsfeld zwischen Bildung und Religion*, ed. Florian Wilk, Themes in Biblical Narrative (Leiden: Brill, forthcoming).

[7] For an analysis of Jub. 4:17–26, see Michael Knibb, "The Book of Enoch in the Light of the Qumran Wisdom Literature," in *Wisdom and Apocalypticism in the Dead Sea Scrolls and the Biblical*

Enoch's writings are considered a source of wisdom, as is reflected in 1QapGen 19:24–25, where Abraham says that because of his own words and his wisdom the noble men from Egypt gave him many gifts: they expect from him "erudition, wisdom, and truth for themselves, so I read before them the book of the words of Enoch." Enoch's role in teaching is explicitly mentioned in 1 En. 81:6 ("We will leave you with your son for one year until (you receive) another order, to teach your children and write for them, and you will testify to all your children").[8] Enoch's wisdom and teaching is closely connected to his ability to write. It is said that he was the first who learned the art of writing, and "who wrote down in a book the signs of the sky in accord with the fixed pattern of their months, so that mankind would know the seasons of the years according to the fixed patterns of each of their month" (Jub. 4:17). He was also "the first to write a testimony" with which "he testified to mankind in the generations of the earth" (Jub. 4:18; cf. 4:19). After the angels had shown him everything on earth and in the heavens, "he wrote down everything. He testified to the watchers" (Jub. 4:21–22). When he was taken from human society and led into the garden of Eden, he wrote about "the judgment and condemnation of the world and all the wickedness of mankind" (Jub. 4:23).[9]

It is also said that Enoch gained his knowledge about cosmology and astronomy from the angels (4:18: "as we [the angels] had told him"; 4:21: "they [God's angels] showed him everything"). He wrote down this knowledge in a book, and with this book he testified against humankind (4:18–19, 23–24) and against the watchers (4:22): he thus related this testimony (4:18) and placed it upon the earth (4:19).

Enoch's book marks the start of a transmission of written knowledge. It is said that Noah received the material from his father Lamech, who received it from his father Methuselah, who in turn received it from his father Enoch (7:38). Noah, in his turn, handed it to his sons (7:39). Noah not only received the knowledge of former generations, he also added his own material: "When he (= Noah) summoned his children, they came to him – they and their children. He divided the earth into the lots, which his three sons would occupy. They reached out their hands and took the book from the bosom of their father Noah" (8:11).

Tradition, ed. Florentino García Martínez, Bibliotheca Ephemeridum Theologicarum Lovaniensium 168 (Leuven: Peeters, 2003), 193–210, esp. 197–99. See also James C. VanderKam, *Enoch: A Man for All Generations* (Columbia: University of South Carolina Press, 1995), 110–21; Van Ruiten, *Primaeval History Interpreted*, 160–66; idem, "A Literary Dependency of *Jubilees* on *1 Enoch*"; Knibb, "Which Parts of *1 Enoch* Were Known to *Jubilees*."

8 See also 1 En. 82:1–2; 83:1; 91:1–3; 94:1.
9 For Enoch as a writer, see also 1 En. 83:2; 13:4,6; 74:2; 82:1.

Like Enoch, Noah was instructed by the angels. One can read that just before he died: "Noah wrote down in a book everything (just) as we [= the angels] had taught him regarding all the kinds of medicine, and the evil spirits were precluded from pursuing Noah's children" (10:13).[10] It was apparently clear that there was more than just one book,[11] the transmission of which occurred through the oldest son: "He gave all the books that he had written to his oldest son Shem because he loved him much more than all his sons" (10:14).

In his turn, Terah taught his son Abraham: "The child [= Abraham] began to realize the errors of the earth – that everyone was going astray after the statues and after impurity. His father taught him (the art of) writing. When he was two weeks of years [= 14 years], he separated from his father in order not to worship idols with him" (11:16–17).[12] He thus realizes the deviations of others and prays to be saved from them. Statues, impurity and wickedness illustrate the errors of the people. It is not said in which language Abraham's father taught him to write, nor the content of this writing. In Jub. 12:25–27 it is said that Hebrew, forgotten after the collapse of Babel, was revived in the days of Abraham "through the revelation

---

10 For an analysis of this passage, see, e.g., James C. VanderKam, "Enoch Traditions in *Jubilees* and Other Second-Century Sources," *Society of Biblical Literature Seminar Papers* 1 (1978): 229–51, esp. 241; Florentino García Martínez, *Qumran and Apocalyptic: Studies on the Aramaic Texts from Qumran*, Studies on the Texts of the Desert of Judah 9 (Leiden: Brill, 1999), 37; Armin Lange, "The Essene Position on Magic and Divination," in *Legal Texts and Legal Issues: Proceedings of the Second Meeting of the International Organization for Qumran Studies: Published in Honour of Joseph M. Baumgarten*, ed. Moshe Bernstein, Florentino García Martínez, and John Kampen, Studies on the Texts of the Desert of Judah 23 (Leiden: Brill, 1997), 377–435, at 383.

11 According to Michael Segal, the "book" (singular) in Jub. 10:13 cannot be identical with the "books" (plural) in Jub. 10:14. The reason for handing over the books to Shem ("because he loved him much more than all his sons") does not match up with the medicines, which were meant for *all* Noah's offspring. The transition from singular to plural even points to separate sources. The nature of the "books" (plural), which Noah handed over to his son Shem, should be understood in the light of the chain of tradition in which knowledge is handed over from generation to generation (Jub. 7:38–39; 12:27; 21:10; 39:6; 45:16). In my opinion, one should not overestimate the transition from singular to plural. The fact that Noah writes a book on medicines (Jub. 10:13) does not exclude the fact that he has written other books. Enoch had written a book (Jub. 4:17–19, 21–23), and in the end of his testament (Jub. 7:38–39), Noah refers to that tradition. It is quite plausible that the new knowledge that Noah received would belong to the chain of tradition. Moreover, it is the offspring of Shem who really need to be protected against the influence of the spirits. The spirits are permitted to only have an influence over other peoples (Jub. 15:31–32; comp. Jub. 10:8). Nevertheless, it is true that the plural in Jub. 10:14 shows that it is not just the knowledge of medicines that is handed over. See Michael Segal, *The Book of Jubilees: Rewritten Bible, Redaction, Ideology and Theology*, Supplements to the Journal for the Study of Judaism 117 (Leiden: Brill, 2007), 171–73.

12 See van Ruiten, *Abraham in the Book of Jubilees*, 25–26.

of an angel." This revelation allowed Abraham to learn about the writings of his forefathers, such as Enoch and Noah (cf. Jub. 21:10). The first thing Abraham does after he learns Hebrew is to copy these books and study them for six months. This means that Abraham, according to Jubilees, had access to esoteric knowledge inherited from the age before Babel, which is often revealed by the angels (e.g. 3:15; 4:15, 18, 21; 10:10–12; cf. also 8:3–4).[13] This would also mean that Abraham's father did not teach him Hebrew and that he had no access to the knowledge of his forefathers before the revelation of the angel.[14] Moreover, it would mean that the books he received from his father (12:27: "his fathers' books") had merely been mechanically handed down, after the collapse, from father to son. Thus, it was not part of his father's instruction, which was instead related to idolatry.

In his testimony to his son Isaac (Jub. 21), Abraham summoned him and gave him orders in relation to offering on the altar.[15] In Jubilees, the patriarchs of the chosen line are priests and they pass along the priestly traditions. Many regulations can be found in the Mosaic legislation in the book of Leviticus, but Jubilees also refers here to the books of his ancestors, with those of Enoch and Noah singled out:

> Eat its meat during that day and on the next day; but the sun is not to set on it on the next day until it is eaten. It is not to be left over for the third day because it is not acceptable to him. For it was not pleasing, and is not therefore commanded. All who eat it will bring guilt on themselves because this is the way I found (it) written in the book of my ancestors, in the words of Enoch and the words of Noah. (21:10)

Isaac's son Jacob also learned the art of writing: "Jacob learned (the art of) writing, but Esau did not learn (it) because he was a rustic man and a hunter" (19:14). Later, Jacob read from the books of his grandfather Abraham to his son Joseph, probably on a regular basis:

> Now Joseph was well formed and very handsome. The wife of his master looked up, saw Joseph, loved him, and pleaded with him to lie with her. But he did not surrender himself. He remembered the Lord and what his father Jacob would read to him from the words of Abraham – that no one is to commit adultery with a woman who has a husband; that there is a death penalty which has been ordained for him in heaven before the most high Lord.

---

[13] Cf. Steven Weitzman, "Why Did the Qumran Community Write in Hebrew?" *Journal of the American Oriental Society* 119 (1999): 35–45.

[14] According to Halpern-Amaru, the sequence of the events (first Terah teaching the skill of writing, then the angel teaching Hebrew) is awkward. It would suggest that Terah knew written Hebrew, but not the spoken language, knowledge of which had been lost at the time of the confusion of tongues. See Halpern-Amaru, *The Perspective from Mt. Sinai*, 58, n. 45.

[15] See van Ruiten, *Abraham in the Book of Jubilees*, 275–93.

> The sin will be entered regarding him in the eternal books forever before the Lord. Joseph remembered what he had said and refused to lie with her. (Jub. 39:5–7)

Joseph resisted the temptations of Potiphar's wife because he knew it would be a sin. He knew this because of his father's teachings.[16]

At the end of his life Jacob gave his books to Levi:

> Israel blessed his sons before he died. He told them everything that would happen to them in the land of Egypt; and he informed them (about) what would happen to them at the end of time. He blessed them and gave Joseph two shares in the land. He slept with his fathers and was buried near his father Abraham in the double cave in the land of Canaan – in the grave, which he had dug for himself in the double cave in the land of Hebron. He gave all his books and the books of his fathers to his son Levi so that he could preserve them and renew them for his sons until today. (Jub. 45:14–16)

Although Exodus makes no mention of any connection between Moses and his Levite father, Jubilees reports that his father Amram taught his son Moses the art of writing: "Afterwards, when you had grown up, you were brought to the pharaoh's daughter and became her child. Your father Amram taught you (the art of) writing. After you had completed three weeks [= 21 years], he brought you into the royal court" (Jub. 47:9). The teaching that Amram imparted to his son puts Moses in the line of all the patriarchs.[17] He is placed within the authoritative written tradition that began with Enoch before the flood and extended through

---

[16] According to James Kugel, this assertion derives from an exegetical motif attached to Gen 49:24: "by the hands of the Mighty One of Jacob." The phrase "the Mighty One of Jacob" (אביר יעקב) was interpreted "as his father Jacob" (אביו יעקב). According to Kugel, the idea that Joseph remembered his father's teachings developed out of this. See also Testament of Joseph 3:3; Joseph and Asenath 7:5; Targum Pseudo-Jonathan 49:24. See James L. Kugel, *A Walk through Jubilees: Studies in the Book of Jubilees and the World of Its Creation*, Supplements to the Journal for the Study of Judaism 156 (Leiden: Brill, 2012), 179–80. See also idem, *The Bible as It Was* (Cambridge: The Belknap Press of Harvard University Press, 1997), 258.

[17] For an analysis of this passage, see Jacques T. A. G. M. van Ruiten, "Moses and His Parents: The Intertextual Relationship between Exodus 1:22–2:10 and Jubilees 47:1–9," in *Rewritten Bible Reconsidered: Proceedings of the Conference in Karkku, Finland, August 24–26, 2006*, ed. Antti Laato and Jacques van Ruiten, Studies in Rewritten Bible 1 (Winona Lake: Eisenbrauns, 2008), 43–78; Halpern-Amaru, *The Perspective from Mt. Sinai*, 49–63. The notion that Moses's father Amram taught him the art of writing seems to be at odds with a tradition that reports on Moses's education in Egyptian wisdom, as we can read in several mostly slightly later sources: *Exagoge* 36–38; Josephus, *Ant.* 2.236. In Philo's *Life of Moses* (1:8–24), Moses was given royal education because his was adopted by the king's daughter. The teaching he received was a combination of Egyptian and Greek elements, and Moses teaches are not only Egyptian, but also Greek, Assyrian, and Chaldean scholars. See Anthony Hilhorst, "'And Moses was Instructed in All the Wisdom of the Egyptians' (Acts 7.22)," in *The Wisdom of Egypt: Jewish, Early Christian and Gnostic Essays*

the patriarchal period to Moses's time. All patriarchs contributed to this written tradition and transmitted it to their favourite sons. The full law would be recorded in Moses' time.[18] At the same time, this also refers to the beginning of the book, where the angels of the presence told Moses to write down all the words (about creation and later events), which then became the very book of Jubilees.[19]

In conclusion, one might say that the narrative of the book of Jubilees presents an image of fathers who educate, generally, their eldest sons by teaching them how to write and by handing down books to them. As we saw, Enoch was the first on earth who learned the art of writing, and he wrote down certain knowledge in a book, and this book was handed down to his sons from generation to generation. This also happened to Noah, Abraham, Isaac, Jacob, Levi and Moses, who gradually added material to this book, and who wrote new books. The knowledge of the fathers is usually said to be derived from the teachings of the angels: Enoch wrote down his testimony as the angels had told him; Noah also wrote down in a book everything just as the angels had taught him; the angel of the presence taught Abraham Hebrew so he would be able to study the books of his fathers; while Joseph himself remembered that heaven had ordained the death penalty for committing adultery.

In this way, education creates a chain of tradition. The purpose of linking knowledge in this chain is, on the one hand, to impart authority and validity to the commandments in the pre-Sinaitic period,[20] and on the other, to give authority to the teaching within Jubilees.[21] However, only part of the knowledge that is to be transmitted is explicitly related to this chain of tradition. Enoch's knowledge was related to the judgment of the children of men, as well as of the watchers, and to the signs of the sky in relation to the calendar; Noah's writings were related to the division of the earth, and to remedies against evil spirits; Abraham taught his son Isaac about the dietary laws; when he was in the house of Potiphar, Joseph remembered Abraham's words that no one is to commit adultery with a woman

---

in Honour of Gerard P. Luttikhuizen, ed. Anthony Hilhorst and George H. van Kooten, Ancient Judaism and Early Christianity 59 (Leiden: Brill, 2005), 153–76, see esp. 162–63.
18 Cf. VanderKam, *Book of Jubilees*, 120.
19 According to Halpern-Amaru, there is no indication that the books of the chain of the tradition came into the possession of Amram or were handed over to Moses. See Halpern-Amaru, *The Perspective from Mt. Sinai*, 58.
20 See Segal, *The Book of Jubilees*, 157.
21 See Hindy Najman, "Interpretation as Primordial Writing: *Jubilees* and Its Authority Conferring Strategies," *Journal for the Study of Judaism* 30 (1999): 379–410 = eadem, *Past Renewals: Interpretative Authority, Renewed Revelation, and the Quest for Perfection in Jewish Antiquity*, Supplements to the Journal for the Study of Judaism 53 (Leiden: Brill, 2010), 39–71.

who has a husband; and, finally, Jacob foretold what would happen in the future, not only the near future, but at the end of time.

By presenting the chain of tradition in Jub. 7:38–39, and Jub. 19:24, Jubilees leaves out certain patriarchs. In the first place, the generations between Shem and Abraham are omitted (Arpachsad, Kainan, Shelah, Eber, Peleg, Ragew, Serug, Nahor, Terah). They are considered erratic, troubled generations.[22] During these generations, the earth was divided, the tower of Babel was built and evil spirits began to have an influence on Noah's grandchildren. As a consequence of the collapse of the tower, the knowledge of the Hebrew language was lost. The antediluvian patriarchs, Kenan and Jared, are also omitted from the chain of tradition.

Nevertheless, even these lost generations receive education in relation to writing. About Kainan it is said that he was instructed in the art of writing by his father, Arpachsad. However, it is not said that he read his father's books. Rather, it is said that he found an inscription on a rock which he could read (8:3), which described the astrological teachings of the watchers "by which they used to observe the omens of the sun, moon, and stars, and every heavenly sign." The teachings of the watchers can be contrasted with the teachings of the patriarchs, which were ultimately received from the good angels.

In the book of Jubilees, the watchers were originally sent to earth by God "to teach mankind and to do what is just and upright on earth" (4:15), but they deviated from the right way.[23] They had illicit intercourse with woman (5:1; 7:21); they committed the first acts of "uncleanness"; and they committed injustice, that is to say, they shed blood (7:23–24). As a result of the fornication, the uncleanness and the injustice, the earth was flooded. The astrological teachings of the watchers are also mentioned in relation to Kainan, who lived after the flood. In the Book of Watchers, the watchers are not only involved in illicit sexual practices and violence, but also in astrological teachings (cf. 1 En. 8:3).

The rejection of astrology is developed further in the book of Jubilees. It also shows, implicitly, that the astrological knowledge taught by the fallen angels, the watchers, is mainly Chaldean in character. Archpachsad, the father of Kainan (who taught him writing), is an ancestor of the Chaldeans (cf. 9:4). Foreign (here especially Chaldean) knowledge is connected with the fallen angels and has to be rejected. However, it is first transmitted in these troubled generations. Nahor, the grandfather of Abraham, was taught this astrology by his father Serug: "He

---

[22] Betsy Halpern-Amaru, *The Empowerment of Women in the Book of* Jubilees, Supplements to the Journal for the Study of Judaism 60 (Leiden: Brill, 1999), 21; Van Ruiten, *Primaeval History Interpreted*, 318.

[23] Van Ruiten, *Primaeval History Interpreted*, 159–60.

[= Nahor] grew up and settled in Ur – in the one that is the Ur of the Chaldeans. His father taught him the studies of Chaldeans: to practice divination and to augur by the signs of the sky" (11:8).

Although this text does not explicitly disapprove of the studies of the Chaldeans, it is striking that in the context of Serug's birth the threats of the evil spirits and their leader Mastema are mentioned (Jub. 11:4–5). The knowledge of the Chaldeans in relation to the practice of divination is closely related to the influence of the evil spirits, which means corruption, destruction and violence. The astrological knowledge of the Chaldeans forms the direct background for Abraham's observations, made to predict the weather for the coming year (Jub. 12:16–18), with which he rejected astrology. Immediately after this, Abraham was called not to go back to Ur of the Chaldeans but to go to the promised land. He was taught Hebrew and started to read the books of his fathers in Hebrew.

3.4 Conclusion

In the book of Jubilees, knowledge is taught by the angels, who learned this from the heavenly tablets. Angels transmitted this supernatural knowledge to distinguished patriarchs (Enoch, Noah, Abraham and Moses). The patriarchs transmitted it to their sons and, ultimately, to all of the children of Israel. The content of this knowledge is of various kinds: halakah (of a certain kind), calendar (especially the solar calendar), knowledge to ensure protection against the influence of evil spirits, and eventually the book of Jubilees itself. What we also see is that certain knowledge may not be transmitted, knowledge which comes from outside Israel and is regarded as a threat to Israel and must therefore be hidden. This concerns the astrological knowledge of the Chaldeans, which is connected to the teachings of the fallen angels: the watchers. In certain generations, forbidden instruction was indeed transmitted after it was found by Kainan on a rock, until Abraham put an end to this.

References

Terence L. Donaldson, *Judaism and the Gentiles: Jewish Patterns of Universalism (to 135 CE)* (Waco: Baylor University Press, 2007).
John C. Endres, *Biblical Interpretation in the Book of Jubilees*, Catholic Biblical Quarterly Monograph Series 18 (Washington: Catholic Biblical Association of America, 1987).
Florentino García Martínez, *Qumran and Apocalyptic: Studies on the Aramaic Texts from Qumran*, Studies on the Texts of the Desert of Judah 9 (Leiden: Brill, 1999).

Martin Goodman, George H. van Kooten, and Jacques T. A. G. M. van Ruiten, eds., *Abraham, the Nations, and the Hagarites: Jewish, Christian, and Islamic Perspectives on Kinship with Abraham*, Themes in Biblical Narrative 13 (Leiden 2010).

Betsy Halpern-Amaru, *The Empowerment of Women in the Book of Jubilees*, Supplements to the Journal for the Study of Judaism 60 (Leiden: Brill, 1999).

Betsy Halpern-Amaru, *The Perspective from Mt. Sinai: The Book of Jubilees and Exodus*, Journal of Ancient Judaism Supplements 21 (Göttingen: Vandenhoeck & Ruprecht, 2015).

Anthony Hilhorst, "'And Moses was Instructed in All the Wisdom of the Egyptians' (Acts 7.22)," in *The Wisdom of Egypt: Jewish, Early Christian and Gnostic Essays in Honour of Gerard P. Luttikhuizen*, ed. Anthony Hilhorst and George H. van Kooten, Ancient Judaism and Early Christianity 59 (Leiden: Brill, 2005), 153–76.

Martha Himmelfarb, *A Kingdom of Priests: Ancestry and Merit in Ancient Judaism* (Philadelphia: University of Pennsylvania Press, 2006).

Michael Knibb, "The Book of Enoch in the Light of the Qumran Wisdom Literature," in *Wisdom and Apocalypticism in the Dead Sea Scrolls and the Biblical Tradition*, ed. Florentino García Martínez, Bibliotheca Ephemeridum Theologicarum Lovaniensium 168 (Leuven: Peeters, 2003), 193–210.

Michael Knibb, "Which Parts of *1 Enoch* Were Known to *Jubilees*? A Note on the Interpretation of *Jub.* 4.16–25," in *Reading from Right to Left: Essays on the Hebrew Bible in Honour of David J.A. Clines*, ed. J. Cheryl Exum and Hugh G. M. Williamson, Journal for the Study of the Old Testament Supplement Series 373 (London: Sheffield Academic Press, 2003), 254–62.

James L. Kugel, *A Walk through Jubilees: Studies in the Book of Jubilees and the World of Its Creation*, Supplements to the Journal for the Study of Judaism 156 (Leiden: Brill, 2012).

James L. Kugel, *The Bible as It Was* (Cambridge: The Belknap Press of Harvard University Press, 1997).

Armin Lange, "The Essene Position on Magic and Divination," in *Legal Texts and Legal Issues: Proceedings of the Second Meeting of the International Organization for Qumran Studies: Published in Honour of Joseph M. Baumgarten*, ed. Moshe Bernstein, Florentino García Martínez, and John Kampen, Studies on the Texts of the Desert of Judah 23 (Leiden: Brill, 1997), 377–435.

Hindy Najman, "Interpretation as Primordial Writing: *Jubilees* and Its Authority Conferring Strategies," *Journal for the Study of Judaism* 30 (1999): 379–410 = eadem, *Past Renewals: Interpretative Authority, Renewed Revelation, and the Quest for Perfection in Jewish Antiquity*, Supplements to the Journal for the Study of Judaism 53 (Leiden: Brill, 2010), 39–71.

Hindy Najman and Eibert Tigchelaar, "Unity after Fragmentation," *Revue de Qumrân* 104/26 (2014): 495–500.

George W. E. Nickelsburg, *Jewish Literature between the Bible and the Mishnah: A Historical and Literary Introduction*, 2nd ed. (Minneapolis: Fortress, 2005).

Jacques T. A. G. M. van Ruiten, "A Literary Dependency of *Jubilees* on *1 Enoch*: A Reassessment of a Thesis of J.C. VanderKam," *Henoch* 26 (2004): 205–9.

Jacques T. A. G. M. van Ruiten, *Primaeval History Interpreted: The Rewriting of Genesis 1–11 in the Book of Jubilees*, Supplements to the Journal for the Study of Judaism 66 (Leiden: Brill, 2000).

Jacques T. A. G. M. van Ruiten, "A Literary Dependency of *Jubilees* on *1 Enoch*: A Reassessment of a Thesis of J.C. VanderKam," *Henoch* 26 (2004): 205–9.

Jacques T. A. G. M. van Ruiten, *Abraham in the Book of Jubilees: The Rewriting of Genesis 11:26–25:10 in the Book of Jubilees 11:14–23:8*, Supplements to the Journal for the Study of Judaism 161 (Leiden: Brill, 2012).

Jacques T. A. G. M. van Ruiten, "Moses and His Parents: The Intertextual Relationship between Exodus 1:22–2:10 and *Jubilees* 47:1–9," in *Rewritten Bible Reconsidered: Proceedings of the Conference in Karkku, Finland, August 24–26, 2006*, ed. Antti Laato and Jacques van Ruiten, Studies in Rewritten Bible 1 (Winona Lake: Eisenbrauns, 2008), 43–78.

Michael Segal, *The Book of Jubilees: Rewritten Bible, Redaction, Ideology and Theology*, Supplements to the Journal for the Study of Judaism 117 (Leiden: Brill, 2007).

James C. VanderKam, "Enoch Traditions in *Jubilees* and Other Second-Century Sources," *Society of Biblical Literature Seminar Papers* 1 (1978): 229–51.

James C. VanderKam, *Enoch: A Man for All Generations* (Columbia: University of South Carolina Press, 1995).

James C. VanderKam, "Recent Scholarship on the Book of *Jubilees*," *Currenst in Research: Biblical Studies* 6 (2008): 405–31.

Steven Weitzman, "Why Did the Qumran Community Write in Hebrew?" *Journal of the American Oriental Society* 119 (1999): 35–45.

Cana Werman, "The Book of *Jubilees* and the Qumran Community: The Relationship between the Two," *Megillot* 2 (2004): 37–55 [Hebrew].

Katell Berthelot
4 The Torah Between Revelation and Concealment in Rabbinic Traditions Pertaining to the Conquest of the Land of Canaan

4.1 Introduction

What religious knowledge is more crucial to the Jews than knowledge of the Torah? In a way, all religious knowledge can be considered Torah, at least according to rabbinic Judaism, which distinguishes between Written and Oral Torah, but views them as deeply united.

If one goes back to the biblical period, when the God of Israel was still conceived of as the God of a specific people as opposed to other peoples with their own "national" gods, the laws of Israel were simply the "national" laws of the Israelites, not a Law given as a Revelation to the entire world. However, after the Exile and with the evolution of Israel's religion toward monotheism, biblical laws started to be considered a universal Law, the Law of the one and only God, the Law of the God of the universe.[1] During the Hellenistic and early Roman period, there were Jewish authors who formulated the idea that the Torah was in agreement and harmony with the Law of Nature, or with the Logos.[2] Knowledge of the Torah could therefore be achieved through the use of one's *logos*, or reason, just as knowledge of God could be achieved through contemplation of the cosmos.

---

[1] The literature on this topic is extensive. See for instance Sidney Greidanus, "The Universal Dimension of Law in the Hebrew Scriptures," *Studies in Religion/Sciences Religieuses* 14/1 (1985): 39–51.

[2] This is particularly clear in the work of Philo of Alexandria. See Carlos Lévy, *Cicero academicus: Recherches sur les "Académiques" et sur la philosophie cicéronienne* (Rome: Ecole française de Rome, 1992), 516–20; John W. Martens, *One God, One Law: Philo of Alexandria on the Mosaic and Greco-Roman Law* (Boston – Leiden: Brill, 2003); Hindy Najman, "A Written Copy of the Law of Nature: An Unthinkable Paradox?" *The Studia Philonica Annual* 15 (2003): 54–63; Gregory E. Sterling, "Universalizing the Particular: Natural Law in Second Temple Jewish Ethics," *The Studia Philonica Annual* 15 (2003): 64–80.

Note: This research has been funded by the European Research Council (ERC) under the European Union's Seventh Framework Program (FP/2007–2013)/ERC Grant Agreement no. 614 424. It has been conducted within the framework of the ERC project JUDAISM AND ROME, under the auspices of the Centre national de la recherche scientifique (CNRS) and Aix-Marseille University, UMR 7297 TDMAM (Aix-en-Provence, France).

This understanding of God and the Torah justified the sharing of religious knowledge between Jews and non-Jews.

Other groups within Judaism, however, saw it differently. The leaders of the communities described in some of the Dead Sea Scrolls could hardly share religious knowledge even with fellow Jews if the latter had not become members of their communities. As for the Palestinian rabbis, they were deeply ambivalent about the possibility of sharing the Torah with non-Jews. In the long run, the view according to which the Torah was given exclusively to Israel and was not meant to be taught to Gentiles or practiced by Gentiles became the majority view in rabbinic Judaism. This evolution was of huge significance for the history of Judaism as a whole.

However, in his book entitled *Torah for the Entire World*, Marc Hirshman has argued that within early rabbinic literature, there were still differing opinions. He has drawn attention to a group of early rabbinic writings that convey what he calls a universalistic understanding of the Torah, according to which it was offered not only to Israel, but to all human beings. According to Hirshman, this universalistic trend is connected to the school of Rabbi Ishmael, a rabbi from the second century CE.[3] A famous example of such a universalistic trend is found in the midrash (or commentary) on the Book of Exodus called Mekhilta deRabbi Ishmael, the final redaction of which dates from the third century CE. In connection with Exodus 19:2, "They (Israel) encamped in the wilderness," the midrash states that:

> The Torah was given publicly (Heb. *dēmos*, from Greek δῆμος), openly (Heb. *parrēsia*, from Greek παρρησία), in a free place. For had the Torah been given in the Land of Israel, (the Israelites) would have said to the nations of the world, 'You have no share in it'.[4] Therefore it was given in the wilderness publicly, openly, in a free place, (so that) everyone wishing to accept it, (could) come and accept it.[5]

As Hirshman notes, "It is striking that the Mekilta not only advanced the claim that revelation had been intended for all peoples, but did so in a vocabulary that

---

[3] See Marc (Menahem) Hirshman, *Torah for the Entire World* (Tel Aviv: Hakibbutz Hameuchad, 1999); idem, "Rabbinic Universalism in the Second and Third Centuries," *Harvard Theological Review* 93 (2000): 101–15.
[4] With a small correction to the edition of Horovitz-Rabin, 205, which has: אין להם ("They do not have a share..."). MS Oxford 151 has אין לכם. This is merely an issue of direct versus indirect speech.
[5] MRI Baḥodesh 1. See Horovitz-Rabin, 205; the translation is based on that of Jacob Z. Lauterbach, *Mekhilta de-Rabbi Ishmael: A Critical Edition, Based on the Manuscripts and Early Editions, with an English Translation, Introduction, and Notes*, 2nd ed. (Philadelphia: The Jewish Publication Society, 2004), 2:293–295, slightly modified. MS Oxford 151 omits the word סומד in the first sentence, but it is found in the second.

was the hallmark of Greco-Roman democracy."⁶ One ought to be more precise, however. The words *dēmos* and *parrēsia* can actually be considered to refer to two different phenomena in the Roman world: on the one hand, the publication of a law *dēmos(ia)*, "in a public place" recalls the norms by which laws, edicts or official letters had to be published in the Roman Empire⁷; on the other hand, the Greek term *parrēsia*, "boldness, freedom of speech," is reminiscent of the Stoic philosophers who opposed the Roman emperors, with the implication that God is put in the role of a philosopher who boldly tells the truth to the world. In both cases, one could not be further away from an expression of religious esotericism.

In the following lines of the midrash, Rabbi Yose emphasizes this point by quoting God's words in Isa 45:19 ("I have not spoken in secret, in the place of a land of darkness" etc.) and by rephrasing the divine speech as follows: "When I gave (the Torah) from the very start, I did not give it in secret, neither in the place of a land of darkness, nor in an obscure place." This means that God did not reveal the Torah to Israel alone, but wanted to share it with all the nations. However, the latter refused to receive it.⁸

Another early rabbinic (Tannaitic) midrash, the Sifre Deuteronomy, declares that God did not reveal the Torah in merely one language, but in four languages, Hebrew, "Roman" (*romi*, meaning Latin), Arabic and Aramean.⁹ Moreover, the midrash goes one step further and states that God did not reveal the Torah to Israel alone, but to all the nations (*umot*). The latter, however, declined God's offer, because they did not want to repent from their wicked ways.

In the end, in both Mekhilta deRabbi Ishmael, traditionally linked to the school of Rabbi Ishmael, and Sifre Deuteronomy, traditionally linked to the school of Rabbi Akiva, the nations are condemned for not having accepted the Torah, and both midrashim foretell their forthcoming judgement.¹⁰

---

6 Hirshman, "Rabbinic Universalism in the Second and Third Centuries," 103.
7 See Katell Berthelot, "Rabbinic Universalism Reconsidered: The Roman Context of some Rabbinic Traditions pertaining to the Revelation of the Torah in Different Languages," forthcoming in *Jewish Quarterly Review* 108.4 (2018). On the Roman norms for the publication of official documents, see Fritz F. Von Schwind, *Zur Frage der Publikation im römischen Recht*, Münchener Beiträge zur Papyrusforschung und Antiken Rechtsgeschichte 31 (Munich: Beck, 1940), esp. 84, 86; Clifford Ando, *Imperial Ideology and Provincial Loyalty in the Roman Empire* (Berkeley: University of California Press, 2000), 80–101.
8 See MRI Baḥodesh 1, ed. Horovitz-Rabin, 206.
9 Sifre Deuteronomy 343 on Deut 33:2; see ed. Finkelstein, 395–97 for the whole argument.
10 See MRI Baḥodesh 1 (after the passage quoted above; ed. Horovitz-Rabin, 206): "R. Eliezer the son of R. Jose the Galilean used to say: Behold it says: *He declares His word to Jacob ... He has not dealt so with any other nation* (Ps 147:19–20). But what had those wretched nations done that He would not give them the Torah? *They do not know his ordinances* (ibid.)—they were unwilling

A straightforward conclusion would seem to be that after the episode at Sinai, the Torah was not meant to be shared with any nation at all, insofar as it had already been offered to the nations of the world, who had unanimously refused it and were now doomed. Other passages in rabbinic literature, however, tell us that the Torah was once again made known to the nations after the Israelites had arrived in Canaan. The idea that the Torah was made known to the Canaanites is highly paradoxical, and contradicts the instructions given to the Israelites in the Bible, which were either to expel or to eradicate the inhabitants of the Land. Nowhere in the Torah is it stated that these nations could repent from their idolatry and their abominations; nowhere are the Israelites supposed to try to convert them to the ways of YHWH. They must simply get rid of them. The rabbinic passages I am alluding to, however, suggest that the Torah was revealed to them and that their fate could have been utterly different. Sharing the Torah with the Canaanites could have led to their inclusion instead of their exclusion.

We shall see that these rabbinic texts actually refer to both the revelation and the concealment of the Torah. In order to understand the logic behind the texts, we first need to go back to the Hebrew Bible and look at a peculiar commandment given by Moses to Joshua and Israel slightly before the conquest. I shall then analyse each rabbinic text and its specific rhetorical strategy separately, in order to show how the argumentation developed from text to text, and to clarify the exegetical, theological and ethical issues at stake in sharing or not sharing knowledge of the Torah with the Gentiles. Finally, I shall try to draw a connection between these texts and the historical context in which they were written, that of the Roman Empire and the attempt by the rabbis to resist imperial domination.

---

to accept them, as it is said: *God comes from Teman ... and a brightness appears as the light ... before Him goes the pestilence ... He stands, and shakes the earth, He beholds, and makes the nations to tremble,* etc. (Hab 3:3–6)" (trans. Lauterbach, *Mekhilta de-Rabbi Ishmael*, 2:295). In Sifre Deuteronomy 343, one reads: "And so (God) asked every nation individually whether they wanted to accept the Torah, as it is said: *All the kings of the earth shall praise you, O Lord, when they hear the words of your mouth* (Ps 138:4). Could it be that they heard and accepted? Scripture says: *And I will execute vengeance in anger and fury upon the nations, such as they have not heard* (Mic 5:14). It was not enough for them not to hear [the Torah], even the seven commandments that the children of Noah accepted [to take] upon themselves they were unable to keep" (ed. Finkelstein, 396; my translation).

4.2 The Biblical Traditions Concerning the Transcription of the Torah Upon the Stones (Deut 27:2–8; Jos 4:1–24; Jos 8:30–34)

The commandments formulated in Deuteronomy 27:2–8 are as follows:

> 2 And on the day you pass over the Jordan to the land which the Lord your God gives you, you shall set up large stones, and plaster them with plaster; 3 and you shall write upon them all the words of this Law (Torah), when you pass over to enter the land which the Lord your God gives you, a land flowing with milk and honey, as the Lord, the God of your fathers, has promised you. 4 And when you have passed over the Jordan, you shall set up these stones, concerning which I command you this day, on Mount Ebal, and you shall plaster them with plaster. 5 And there you shall build an altar to the Lord your God, an altar of stones: you shall lift up no iron tool upon them. 6 You shall build an altar to the Lord your God of unhewn stones; and you shall offer burnt offerings on it to the Lord your God; 7 and you shall sacrifice peace offerings, and shall eat there; and you shall rejoice before the Lord your God. 8 And you shall write upon the stones all the words of this law *very plainly* (*ba'er heytev*) (trans. NRSV).

The wording of this text is highly ambiguous. Verses 2 and 4 seem to repeat each other: one wonders whether large stones are to be erected twice, first immediately after the crossing of the Jordan and then again on Mount Ebal, or whether the text refers to the same stones. If the text refers to the same stones, are they to be plastered twice, and, in this case, would the second plaster not cover the words of the Law? Finally, what about the stones of the altar? Does verse 8 mean that the Law is to be written upon the stones of the altar as well? Or is it simply summarizing the commandment given in this passage? In short, this text is replete with ambiguities.

The account of the Book of Joshua, which supposedly describes how the conquest happened, clarifies these issues only to a certain extent. Chapter 4 tells the story of the miraculous crossing of the Jordan River by Israel. The Israelites pick up twelve stones in the bed of the river and Joshua has the stones (steles) erected at Gilgal, as a memorial for future generations, who will thus remember the miracle performed by God for their ancestors, and as a testimony to all the peoples of the world (Josh 4:24). However, these stones are not plastered with plaster and can hardly be identified with the stones mentioned in Deuteronomy 27. In the Book of Joshua, the commandment of Deuteronomy 27 is performed at a later stage, on Mount Ebal, together with the construction of the altar (Joshua 8:30–35).[11]

---

11 "30 Then Joshua built an altar in mount Ebal to the Lord, the God of Israel, 31 as Moses the servant of the Lord had commanded the people of Israel, as it is written in the book of the

We read in Joshua 8:32: "And there, in the presence of the people of Israel, he wrote upon the stones a copy of the law of Moses which he had written" (trans. NRSV). It seems that in this case, the stones on which the Torah was written are those of the altar, mentioned in the previous verses. But there is no reference to the plaster. Thus, when all the biblical passages related to the stones, the transcription of the Torah and the memorial are taken together, some contradictions or unclear points remain.

4.3 Mishnah Soṭah 7:5

In rabbinic literature, the story of the stones on which the Torah was written is first referred to briefly in the Mishnah, tractate Soṭah, at the end of a section that deals with the blessings and curses to be pronounced by Israel on Mount Garizim and Mount Ebal (Deut 27–28, Josh 8:30–35):

> And afterward they brought the stones and built the altar and plastered them [the stones] with plaster. And they wrote upon them all the words of this Law [in] seventy languages, as it is written, *very plainly* (*ba'er heyṭev*; Deut 27:8). And they took the stones and came and spent the night in their own place.[12] (M. Soṭah 7:5)

The Mishnah seems to imply that the Torah was written on the stones of the altar, as in the Book of Joshua. The biblical passage referred to, however, is Deuteronomy 27:8, while the idea that the Israelites brought the stones (those taken from the midst of the Jordan) in the place where they were about to spend the night comes from Joshua 4:3 and 8 (the context of which refers to the stones taken from the midst of the Jordan river).

---

law of Moses, 'an altar of unhewn stones, upon which no man has lifted an iron tool'; and they offered on it burnt offerings to the Lord, and sacrificed peace offerings. 32 And there, in the presence of the people of Israel, he wrote upon the stones a copy of the law of Moses, which he had written. 33 And all Israel, sojourner as well as homeborn, with their elders and officers and their judges, stood on opposite sides of the ark before the Levitical priests who carried the ark of the covenant of the Lord, half of them in front of Mount Gerizim and half of them in front of Mount Ebal, as Moses the servant of the Lord had commanded at the first, that they should bless the people of Israel. 34 And afterward he read all the words of the law, the blessing and the curse, according to all that is written in the book of the law. 35 There was not a word of all that Moses commanded which Joshua did not read before all the assembly of Israel, and the women, and the little ones, and the sojourners who lived among them" (trans. NRSV, 272–73).

12 MS Kaufmann, the translation is mine.

The context of this mishnah is a discussion starting in Mishnah Soṭah 7:1 about the prayers or texts that may be recited in any language and those that must be recited in Hebrew.[13] Among the latter are "the paragraph of the first-fruits, the words of *ḥalitzah*, the Blessings and the Cursings [the passage immediately following the one concerning the stones, Deut 27:12–28:68], the Blessings of the priests and the Blessings of the high priest, the paragraph of the king, the paragraph of the heifer whose neck is to be broken, and [the words of] the Anointed for battle when he speaks unto the people."[14] This list is based on a typically rabbinic hermeneutical principle, according to which cases mentioned in different biblical verses are associated with each other because these verses share a similar formulation. In this mishnah, the essential feature is the combination of the verbs "answer" (*'anah*) and "say" (*'amar*) in the implied biblical verses, a combination that looks quite repetitive and superfluous.

Mishnah Soṭah 7:1–5 thus states that the Blessings and Cursings in chapters 27–28 of Deuteronomy are to be recited in Hebrew, but on the other hand, building upon Deuteronomy 27:8, it understands the expression *ba'er heyṭev*, "very plainly," as referring to the translation of the Torah in seventy languages. What does it mean? The number seventy is highly symbolic and means that the Torah was translated into all the languages of humankind. It may imply that the Torah was communicated to the nations, but not necessarily so. As both Willem Smelik and Steven Fraade have argued, if one looks at this mishnaic passage independently of the traditions found in the Tosefta and later rabbinic texts, the Mishnah's very brief statement can be understood to reflect a philosophy of language according to which the Torah had to be formulated in all the languages of the world in order for all the nuances of its meaning to be manifested. According to such an interpretation, the translation of the Torah was meant only for the Jews, or – if one follows Smelik's analysis – was not meant for anyone in particular.[15]

---

13 On this section of the Mishnah, see Judith Hauptman, *Rereading the Mishnah: A New Approach to Ancient Jewish Texts*, Texts and Studies in Ancient Judaism 109 (Tübingen: Mohr Siebeck, 2005), 196–207. In this case she sees the Mishnah as a reworking of the parallel tradition in the Tosefta.
14 Trans. by Herbert Danby, *The Mishnah* (Oxford: Oxford University Press, 1933), 300.
15 See Willem Smelik, *Rabbis, Language and Translation in Late Antiquity* (Cambridge: Cambridge University Press, 2013), 29–30, who writes: "The Mishnah does not indicate that these translations were aimed at the nations—not even that they were aimed at Israel's enlightenment." See also (with some differences) Steven Fraade, "Before and After Babel: Linguistic Exceptionalism and Pluralism in Early Rabbinic Literature and Jewish Antiquity," *Diné Israel* 28 (2011): 31*–68*, esp. 54*–55*; idem, "The Torah Inscribed/Transcribed in Seventy Languages," in *Hebrew between Jews and Christians*, ed. Daniel Stein Kokin, Studia Judaica (Berlin: De

Whereas this interpretation is definitely a possibility, the other interpretation, that the translation was meant to communicate the Torah to the nations, also remains possible. In this case, the underlying idea would be that the Torah was made known to *all* the nations of the world, including the Canaanites. This point is reminiscent of the tradition found in the passage of the Mekhilta deRabbi Ishmael referred to in the introduction, and this mishnah has thus been associated with the school of Rabbi Ishmael.[16] The Mishnah, however, does not clarify what the publication of the Torah in seventy languages implied for the Israelites' encounter with the Canaanites. We must look at the Tosefta for a slightly more detailed account.

4.4 Tosefta Soṭah 8:6–7

The eighth chapter of tractate Soṭah in the Tosefta deals with the crossing of the Jordan river by Israel on their way to the Land of Canaan, the instructions given to Israel by Joshua, the miracle of the waters that were cut off before the ark of the covenant, the twelve stones that were set up in the midst of the Jordan in the place where the feet of the priests who bear the ark stood (Josh 4:5), and the stones that were carried away from the midst of the Jordan (Joshua 4:3, 8, 20). In §§ 8:6–7, we encounter a different version of the tradition found in Mishnah Soṭah 7:5:

> R. Yehudah says: They wrote it [the Torah] on the stones of [the] altar. They told him: How did the nations of the world learn [the laws of] the Torah? He told them: This teaches that God inspired the heart of every nation and every kingdom, and they sent their scribes (*notarim*, from *notarius* in Latin), and they transcribed the text that stood on the back of the stones in seventy languages. On that very hour, the decree of the judgement of the nations of the world was sealed, [and they were doomed] to the pit of destruction.
>
> R. Shimeon says: They wrote [the laws of the Torah] on the plaster. How? They panelled it [i.e., the altar] and plastered it with plaster, and they wrote on it all the words of the Torah in seventy languages. And at the bottom they wrote: *That they may not teach them* [or: *you*, as in MT] (*to do according to all their abominable practices which they have done in the service of their gods*) (Deut 20:18). If you repent from them,[17] we shall receive [or: accept] you.[18] (T. Soṭah 8:6–7)

Gruyter, forthcoming). Fraade considers that the translation was meant for Israel; the underlying implication could be that no matter where Jews would live and what language they would speak, access to the Torah would be possible for them.

16 See Hirshman, *Torah for the Entire World*, 110, 113. And see below for the analysis of the passage from Mekhilta Deuteronomy.

17 With a small correction (בהם instead of בכם).

18 The translation is mine and is based on the Hebrew text of MS Vienna 46, consulted on the website of the Academy of the Hebrew Language (Ma'agarim) (see also ed. Lieberman, 205,

Whereas the tradition in the Mishnah was anonymous, here we have a discussion between two tannaim who lived in the second century CE, Rabbi Yehudah and Rabbi Shimeon.[19] Both seem to rely on a tradition connected to the one found in the Mishnah, since both mention the seventy languages, but the nature of the relationship between the Mishnah and the Tosefta is unclear in this case. They could both rely on a common, previous tradition.[20]

R. Yehudah and R. Shimeon understand the biblical texts in different ways. The first point of divergence between them has to do with the technical details of the act of copying: Was the Torah written directly on the stones of the altar or on the plaster laid on the stones, which would have made the text much easier to read? While R. Yehudah defends the first option, R. Shimeon opts for the second.[21] R. Yehudah seems to rely on a literal reading of Deuteronomy 27:8, "And you shall write upon the stones," whereas R. Shimeon seems to follow the wording and the sequence of actions found in Deuteronomy 27:2–3 – "you shall set up large stones, and plaster them with plaster, and you shall write upon them" – , which also seems to be the Mishnah's source of inspiration: "… they plastered them (the stones) with plaster. And they wrote upon them all the words of this Law …."

A second divergence between the two sages may have to do with the language (or languages) in which the Torah was written on the stones. The statement of

---

which has a slightly different text; the overall meaning, however, remains the same):

ר' יהודה אום 'על אבני מזבח כתבוה. אמרו לו היאך למדו אותן אומות העולם את התורה. אמ' להן מלמד שנתן המקום בלב כל אומה ומלכות ושלחו נטורים שלהם והשיאו את הכתב מגבי אבנים בשבעים לשון. באותה שעה נתחתם גזר דינם של אומות העולם לבאר שחת.

ר' שמעון או 'על הסיד כתבו. כיצד. כירוהו וסדוהו בסיד וכתבו עליו את כל דברי התורה בשבעים לשון וכתבו מלמטה למען אשר לא ילמדו אתם וגו'. 'אם אתם חוזרין בכם אנו מקבלין אתכם.

19 On R. Shimeon (ben or bar Yohai) and R. Yehudah (bar Ilai), who were students of R. Akiva, see Herman L. Strack and Günter Stemberger, *Introduction to the Talmud and Midrash* (Minneapolis: Fortress, 1996), 76–77.

20 This is the position of Steven Fraade (see the bibliographical references in note 15 above, especially "Before and After Babel," 55*). Judith Hauptman considers that "the redactor of the Mishnah reworked the Tosefta in order to make a number of points of his own," so that the mishnaic text represents a shortened version of T. Soṭah 8:6–9 (*Rereading the Mishnah*, 109–24, quotation at 116). Smelik rejects Hauptman's theory (*Rabbis, Language and Translation*, 32). On the problem of the relationship between the Mishnah and the Tosefta in a more general way, see Shamma Friedman, "The Primacy of Tosefta to Mishnah in Synoptic Parallels," in *Introducing Tosefta: Textual, Intratextual and Intertextual Studies*, ed. Harry Fox and Tirzah Meacham (Hoboken: Ktav, 1999), 99–121.

21 This discussion is slightly more developed in the Babylonian Talmud (Soṭah 35b–36a); see below, and see the discussion in Saul Lieberman, *Tosefta Ki-Fshutah: Nashim*, 3 (New York: The Jewish Theological Seminary, 1995), 699–702.

R. Shimeon is close to the concise wording of the Mishnah, and makes clear that it was the Israelites themselves who transcribed the laws of the Torah in seventy languages (how they had the necessary knowledge to do so remains a mystery, but some divine revelation is probably presupposed). On the other hand, the formulation of R. Yehudah's opinion is ambiguous and could be understood to mean that the Torah was written in Hebrew, and then transcribed in the different languages of the world by the scribes of the nations themselves, who were divinely inspired to translate the text on the spot, each into his own language. Alternately, R. Yehudah could have meant that the Torah was translated into seventy languages and copied on the stones by the Israelites, and that the scribes merely transcribed the text written in their own language.[22] In this case he would not differ (on this point) from R. Shimeon.

Finally, and most importantly, the purpose of the "publication" of the Torah and of the translation into seventy languages differs greatly from sage to sage. The rationale underlying R. Yehudah's interpretation is that official knowledge of the Torah was given to the nations so that the nations could be justly condemned.[23] A very important rabbinic adage states that you cannot condemn someone if he has not first been informed and warned. Hence the importance of communicating the commandments of God to the Gentiles, in order for their guilt to be clearly established.[24] Rabbi Shimeon, on the contrary, leaves the possibility of the Gentiles' repentance open. This is clearly stated at the end of

---

[22] The use of the verb השיאו ("transcribed," lit. "lifted," a *hiphil* form of נשא) seems to support the second interpretation, but both are possible. See Marcus Jastrow, *A Dictionary of the Targum, the Talmud Babli and Jerushalmi, and the Midrashic Literature* (New York: Choreb, 1926), 938, who quotes this passage.

[23] There seems to be a pun between the expression in Deut 27:8, "very plainly," *ba'er heyṭev*, which the Mishnah understands as referring to the seventy languages, and the expression used by R. Yehudah who says that as soon as the nations learned about the Torah they were doomed to the "pit of destruction," *be'er šaḥat* (mentioned only once in the Bible, in Ps 55:24). The words *ba'er* and *be'er* look exactly the same.

[24] See Hirshman, *Torah for the Entire World*, 106–7. Saul Lieberman recalls that according to a late midrash, the children of Israel could be punished for their transgressions only after the Torah had been officially published in the Tent of Meeting, and not immediately after the revelation at Sinai. He writes that some rabbinic texts "argued according to the legal practice of the Roman government. An edict had to be displayed δημοσία, in a public place; until then the people were not punishable for its transgression. Similarly, some Rabbis maintained, the Gentiles were not punishable for the transgression of the Torah until it was inscribed on the stones by Joshua. It is by virtue of the publication of the Torah on those στῆλαι, that the Gentiles received their death sentence ... for its transgression" ("Appendix II: The Publication of the Torah," in *Hellenism in Jewish Palestine*, 2nd ed. [New York: The Jewish Theological Seminary, 1994], 201).

the passage, which is directed at the nations: "If you repent from them (that is, from your idolatrous and immoral practices), we shall receive (or: accept) you." Therefore the translation of the Torah has, at least theoretically or ideally, a different purpose than it does in the interpretation of R. Yehudah: it creates a real opportunity for the "conversion" of the Gentiles, at least in the sense of giving up idolatry.

Rabbi Shimeon specifies that at the bottom of the altar, the Israelites wrote: "That they may not teach them [or: you, as in the Massoretic text] (to do according to all their abominable practices which they have done in the service of their gods)". This is a quotation from Deut 20:18. Deuteronomy 20 deals with the laws of war, the topic discussed at the beginning of chapter 7 of Tosefta Soṭah. According to Deut 20:10–18, there is a clear difference between the wars led by the Israelites against "the cities which are very far from you," outside the Land of Canaan, to which peace can be offered (Deut 20:10–15), and those waged against "the cities of these peoples that the Lord your God gives you for an inheritance" (Deut 20:16), which must be utterly destroyed. Rabbi Shimeon's reference to a commandment pertaining specifically to the Canaanites means that the translation of the Torah was also directed at them, that they were warned of their future fate, and that they were given the possibility of repenting and being saved, instead of being put to the ban (ḥerem). This contradicts the literal meaning of biblical passages such as Deut 20, but is indisputable in view of the sentence "If you repent from them, we shall receive you."

The view expressed by Rabbi Shimeon solves a possible ethical dilemma in connection with the Canaanites. Since, according to biblical data, they had not attacked the Israelites first, and insofar as they had not been warned that they should give up their abominable practices, their removal or destruction could seem unfair.[25] The publication of the Torah in the languages of the nations, including that of the Canaanites, was meant to make them aware of the necessity of repenting from their evil ways. Once formally warned, the Canaanites had a choice, and could blame only themselves if their stubborness led to their destruction. This is, at least, the underlying logic behind this passage.[26]

---

25 See, for a comparison, how the author of the Wisdom of Salomon deals with the Canaanites (11:23–12:11). In this context too, the main issue is to justify the divine decision to dispossess and destroy them.
26 Other rabbinic texts emphasize that the Canaanites were given a choice: they could make peace with Israel, or leave, or wage war. See in particular y. Shevi'it 6:1, 36c; Leviticus Rabbah 17:5–6. The idea that the Canaanites could have made peace with Israel is also found in the passage from Mekhilta Deuteronomy examined below (§4); see in particular note 33.

4.5 Mekhilta Deuteronomy on Deuteronomy 27:8

We find a similar logic at work in yet another tannaitic text pertaining to the issue of the Torah written on the stones after Israel's entrance into Canaan. The Mekhilta Deuteronomy is a midrash usually associated with the school of Rabbi Ishmael, which is preserved only in fragments from the Cairo Genizah or in late compilations which have added other elements to the original midrash, making it difficult to recover. Here I rely on a fragment from the Genizah and follow the edition of Menahem Kahana[27]:

> 5 (...) On that very day Israel crossed (the Jordan), and they took the stones and carried them away 6 and erected them and they wrote on [the stones] all the words of the Torah [in the holy language (i.e., Hebrew)]. 7 R. Ishmael says: they wrote (them) in seventy languages [as it is said: *very plainly* (Deut 27:8)]. Rabbi 8 Shimeon ben Yohai says: They did not write (anything) upon the[m b]ut[a copy of] the Law (Torah) of Moses, as it is said: 9 *He wrote there, upon the stones, a copy of the law of Moses* (etc.) (Josh 8:32). vacat R. Yose 10 ben Yosi[28] says in the name of R. Eleazar ben Shimeon: They did not write upon them (anything) but what the nations 11 of the world want; for example: *When you draw near to a city to fight against it, offer terms of peace to it. 12 If its answer to you is peace (and it opens to you, then all the people who are found in it shall do forced labor for you and shall serve you)* (Deut 20:10–11); *When you besiege a city for a long time, (making war against it in order to take it, you shall not destroy its trees* etc.) (Deut 20:19). (It is) on [the stones] 13 [of the alta]r that they wrote them (the words of Torah), according to R. Yehudah. R. Shimeon said: They wrote them upon the stones (which were set up on Mount Ebal). [R. ... said:] 14 The words of R. Shimeon who said "[They wrote them] upon the stones" look (more convincing) 15 [– as it is said: *Upon*] *the stones* (Josh 8:32) – than the words of R. Yehudah who said "They wrote them upon the

---

27 The translation is mine. For the Hebrew text, see Menahem Kahana, *The Genizah Fragments of the Halakhic Midrashim. Part 1* (Jerusalem: The Hebrew University Magnes Press, 2005), 345, n°10, l.5–17:

5 (...) בו ביום עברו ישראל ונטלו את האבנים והעבירום
6 והעמידום וכתבו על [האבנים] את כל דברי התורה [בלשון הקודש]
7 ר' ישמעאל אומ' בשבעים לשון כתבו [שנ' באר היטב] רבי
8 שמעון בן יוחאי א' לא כתבו עליה[ן א]ל[א את משנה] תורת מושה שנ'
9 ויכתב שם על האבנים את משנה תורת מושה וג'
10 בן יוסי אומ' משום ר' אלעזר בן שמעון לא כתבו עליהן אלא מה שאומות
11 העולם רוצין כגון כי תקרב אל עיר להלחם עליה וקראת עליה לשלום
12 אם שלום תענך וג' כי תצור אל עיר ימים רבים וג' על [אבני]
13 [המזב]ח כתבום דברי ר' יודה ר' שמעון א' על האבנים כתבום [אמ']
14 [ר' נרא]ין דברי ר' שמעון שאמר על האבנים [כתבום]
15 [שנ' על] האבנים מדברי ר' יודה שאמר על המזבח כתבום שאלו
16 [על] המזבח כתבום האיך היו אומות העולם רוצין לקרות דין
17 [ולמטה כת'] עליהם כל הרוצה לקבל ימין יבוא ויקבל וגנזום בו ביום.

See also Lieberman, *Tosefta Ki-Fshutah. Nashim. 3*, 700–701; Hirshman, *Torah for the Entire World*, 109–10; Fraade, "The Torah Inscribed/Transcribed in Seventy Languages."
28 Lieberman notes that this is the only reference to a tannaitic sage named in such a way (*Tosefta Ki-Fshutah: Nashim, 3*, 700, n. 17).

altar." They asked: 16 They wrote them [upon] the altar? How (could) the nations of the world see[29] their judgement happen? 17 [And at the bottom they wrote] on them: "Everyone who wants to make peace (lit.: receive the right [hand]), let him come and make peace." And they hid them (the stones) on that very day.

This text shares several features with the passages from the Mishnah and the Tosefta that were examined earlier. It echoes the tradition found in the Mishnah that the Torah was written on the stones in seventy languages, and connects it with Rabbi Ishmael, known for his universalistic views.[30] It also reproduces the discussion between Rabbi Yehudah and Rabbi Shimeon found in the Tosefta, but has a different version of it.

The argument is as follows: the first issue discussed by the Mekhilta, if one follows Kahana's reconstruction of the text in line 6, is language. Then comes the issue of content; and, finally, the question of whether the Torah was written on the stones of the altar or on the stones which were set up on Mount Ebal (Deut 27:4).

Concerning the first point, the issue of language, the opinion that the Torah was written on the stones in Hebrew (according to Kahana's reconstruction) is opposed by Rabbi Ishmael, who, in agreement with the Mishnah, states that the Torah was written on the stones in seventy languages. The fact that this potentially universalistic idea is attributed to Rabbi Ishmael makes sense in view of Marc Hirshman's characterization of Rabbi Ishmael's school. Rabbi Shimeon ben Yohai then argues that they wrote "a copy of the Law of Moses," *mishne Torat Moshe* (a reference to Joshua 8:32). This statement seems to imply that the Law of Moses was copied only in Hebrew, and not in other languages. If this reading is correct, the Mekhilta attributes to R. Shimeon a view quite opposed to the one associated with him in the Tosefta, but this would make sense coming from a disciple of Rabbi Akiva, whose positions are characterized as less universalistic than those of R. Ishmael.

After Rabbi Shimeon ben Yohai's statement comes a *vacat*, which may indicate that the discussion enters into a new stage. However, the repetition of the same phrase or wording by Rabbi Shimeon and Rabbi Yose, and the nature of the arguments, clearly show that their statements are connected. As a matter of fact, Rabbi Yose answers Rabbi Shimeon ben Yohai, but takes the discussion in a new direction, debating not merely the issue of language but the issue of

---

29 Following a correction proposed by Saul Lieberman in *Tosefta Ki-Fshutah: Nashim, 3*, 701 (רואין instead of רוצין). One could also understand רוצין, "want," as referring to the choice the nations had to make once the divine laws were communicated to them. Still, the syntax would be odd.

30 See Hirshman, "Rabbinic Universalism"; idem, *Torah for the Entire World*, 109–10.

content – which part of the Torah was written on the stones?[31] In the first stage of the discussion it was stated that "all the words of the Torah" were written on the stones. According to R. Shimeon, they wrote "a copy of the Law of Moses," and Marc Hirshman interprets this as a reference to the Book of Deuteronomy alone, which is indeed a repetition or a reformulation (*mishne*) of the Law of Moses. For Rabbi Yose, however, it was not the entire Book of Deuteronomy that was written on the stones, but merely the commandments pertaining to the wars between Israel and the nations. The first quotation, Deut 20:10–11, refers to the rules for a war led by Israel against cities located outside Canaan, the only ones to which peace can be offered. The second quotation, Deut 20:19, is also a general rule of war, pertaining to trees: unlike human beings, trees are not enemies and should not be destroyed. One could thus conclude that Israel wrote down only the commandments pertaining to the wars against the nations that were not from Canaan. The end of the text, however, shows that the midrash actually did not follow the literal meaning of Deuteronomy 20, and understood that the offer of peace should be extended to the Canaanites as well. Therefore it is written in line 17: "Everyone who wants to make peace, let him come and make peace." "Everyone" includes the Canaanites. This radical re-interpretation of Deuteronomy 20 can be found in other rabbinic texts as well, for instance in the Jerusalem Talmud, Sheviʻit 6.1 (36c), in Leviticus Rabbah 17.5–6 (on Lev 14:34) or in Deuteronomy Rabbah 5.14 (on Deut 20:10), which all state that before the conquest of Canaan, Joshua sent *prostagmata* and offered peace to the Canaanites who wanted to make peace.[32]

---

[31] Note that in the parallel discussions in the Jerusalem Talmud (Soṭah 7.5 [21d]) and in the Babylonian Talmud (Soṭah 35b–36a), the issue of the content of the Torah written on the stones is not debated. This point is specific to Mekhilta Deuteronomy.

[32] On these rabbinic traditions, see Wilhelm Bacher, "The Supposed Inscription upon 'Joshua the Robber,' Illustrated from Jewish Sources," *Jewish Quarterly Review* 3 (1891): 354–55; Victor Aptowitzer, "Les premiers possesseurs de Canaan, légendes apologétiques et exégétiques," *Revue des études juives* 82 (1926): 274–86; Hans (Yohanan) Lewy, "Ein Rechtsstreit um Boden Palästinas im Altertum," *Monatsschrift für Geschichte und Wissenschaft des Judentums* 77 (1933): 84–99, 172–80; Philip S. Alexander, "The Toponymy of the Targumim, with Special Reference to the Table of the Nations and the Boundaries of the Land of Israel" (PhD diss., University of Oxford, 1974), 92–105; Katell Berthelot, "The Canaanites who 'trusted in God': an original interpretation of the fate of the Canaanites in rabbinic literature," *Journal of Jewish Studies* 62 (2011): 233–61; Menahem Kister, "The Fate of the Canaanites and the Despoliation of the Egyptians. Polemics among Jews, Pagans, Christians, and Gnostics: Motifs and Motives," in *The Gift of the Land and the Fate of the Canaanites in Jewish Thought*, ed. Katell Berthelot, Joseph David, and Marc Hirshman (New York: Oxford University Press, 2014), 66–111. Moshe Weinfeld thought that the verses in Deut 20:10–14 were actually the original version of the tradition pertaining to Israel's interaction with the Canaanites, and that this version was reformulated later on. See Moshe Weinfeld, "The Ban on the Canaanites in the Biblical Codes," in *History and Traditions of Early*

In the end, the Mekhilta argues that the possibility of making peace with Israel was communicated to the Canaanites through the revelation of at least these crucial verses of the Torah. The tragic fate of the Canaanites was the result of their rejection of the peace offered to them. The conclusion is thus similar to that of the Tosefta, except that here the focus is on peace and not on repentance.[33] That the underlying question of the midrash is whether the Canaanites's fate was just or not, and therefore whether they had been warned or not, is shown by the question raised in line 17: "How (could) the nations of the world see their judgement happen?" In other words: How could they know and choose, if they were not told about the possibility of making peace?

This underlying question leads to the last issue discussed in the Mekhilta, the nature of the stones on which the Torah (or part of it) was supposed to be written. Rabbi Yehudah argues that they wrote on the stones of the altar, whereas Rabbi Shimeon seems to think that they wrote on the stones that were set up on Mount Ebal (the text is quite elliptical at these lines). In any case, the final remark that "they hid them (the stones) on that very day" makes sense only if it refers to the stones of the altar. The altar built on Mount Ebal was a temporary one, meant to be demolished after the Israelites' departure; and the stones which had been used for the altar were supposed to be hidden.[34] Therefore, the problem with the fact that the words of the Torah were written on the stones of the altar was that they were displayed for a very short time only. Hence the question in line 17: "How (could) the nations of the world see their judgement happen?" (meaning: if the altar was to be demolished shortly afterwards). In the Tosefta, Rabbi Yehudah answered this question by saying that God had inspired the nations, which had sent envoys to transcribe the Torah that was written on the stones (an answer reproduced in the corresponding passage of the Jerusalem Talmud). Here this tradition seems to be implicitly presupposed, unless one wants to read the last sentence, "And they hid them (the stones) on that very day," in a cynical way, as if the information was formally published, but in such a way as to make it impossible for the nations to become aware of it. Whatever the solution imagined by the rabbis, the main point here is that the apparently superfluous discussion about the nature of the stones on which the Torah was written has to do with the underlying issue dealt with by this text

---

Israel, ed. André Lemaire, Supplements to Vetus Testamentum 50 (Leiden: Brill, 1993), 142–60, see 152–53; idem, *The Promise of the Land: The Inheritance of the Land of Canaan by the Israelites* (Berkeley: University of California Press, 1993), 96–97.

33 Peace probably implied repentance from idolatry as well, at least if the Canaanites were to stay in the country.

34 This is made clear in the discussion found in the Jerusalem Talmud, Soṭah 7.5 (21d).

(and the passage in the Tosefta), i.e., the justification for the punishment of the nations, and especially of the Canaanites.

4.6 The Continuation of the Debate in the Two Talmuds

The discussion of the Torah written on the stones and revealed to the nations continues in the two Talmuds. The Jerusalem Talmud (Soṭah 7.5 [21d]) first raises the question of the identification of the stones on which the Torah was copied in seventy languages: were they the stones that were brought to the place where the Israelites were about to spend the night – the stones "of the lodging place" – or, rather, the stones of the altar? As mentioned previously, the problem raised by the second possibility is that this altar was not permanent; it was meant to be dismantled, and the stones were supposed to be hidden. Therefore the question arises: in such a case, how could the nations learn the Law? The answer, which in the Jerusalem Talmud is anonymous, is that God performed a miracle and "gave insight into the heart of every nation, so that they transcribed the Torah, which was written in seventy languages" (an answer similar to that of R. Yehudah in the Tosefta). Then comes the question of the plaster (*sid*): whereas it is easy to imagine that the stones set up as a memorial would be plastered before the Torah would be written upon them, with the altar the question arises as to whether the plaster would cover the words written on the stones. The answer, however, is that the plaster was laid only between the stones of the altar, and not on the stones themselves. Finally, the discussion concludes with a *gezera shava*, an exegetical technique that brings together two verses on the basis of a word or an expression that they have in common. The plaster (*sid*) referred to in Deut 27:2 and 4 is connected with the plaster or lime (*sid*) mentioned in Isa 33:12, which states that "the peoples will be as if burned to lime." In accordance with another famous exegetical principle, that of "measure for measure," the text states that because the nations refused to obey the commandments of the Torah written on the plaster, they were doomed to be burnt and turned into lime (plaster): in other terms, they will be punished in a way that recalls the nature of their sin. The passage ends with an additional "measure for measure" argument based on another biblical quotation. "The nations shall be utterly laid waste (*ḥarov yeḥeravu*)" (Isa 60:12) leads to the statement that "from Ḥorev comes their sentence to death": the nations shall be destroyed because they have rejected the precepts of the Torah given at Sinai/Ḥorev.

The discussion in the Jerusalem Talmud echoes the one found in the Tosefta and in Mekhilta Deuteronomy, except that here the opinion that the Torah was written on the altar is attributed to R. Yose and not to R. Yehudah. As shown above, this opinion is based on the ambiguity of Deut 27 and on the wording of the Mishnah, which states that the Torah was written on the plastered stones of the altar in seventy languages. However, since the Mishnah itself also mentions the stones that were carried away to the place where the Israelites were about to spend the night, the identification of the stones comes into question. In the end, according to the discussion in the Jerusalem Talmud, either the Torah was written on the stones of the lodging place, in which case it was made easily accessible to the nations and remained so even after the departure of the children of Israel, or the Torah was made known to the nations only for a short time on the stones of the altar, making a miracle necessary in order for the nations to become aware of the Law before the latter was hidden from them. In both cases, however, the nations have had access to the content of the Law, and their refusal to live according to the precepts of the Torah shall lead to their condemnation.

Finally, in the Babylonian Talmud (Soṭah 35b–36a), the discussion – which again takes place between R. Yehudah and R. Shimeon, as in the Tosefta and Mekhilta Deuteronomy – revolves around the question: was the Torah written directly on the stones or on the plaster?[35] According to a literal understanding of Deut 27:3, "you shall write upon them (the stones) all the words of this Law," R. Yehudah argues that they wrote the Torah upon the stones, and plastered them with plaster only afterwards (in accordance with Deut 27:4). But then the question arises: how could the nations know the Torah if the words were covered by the plaster? R. Yehudah answers that God inspired the nations, who sent scribes who peeled off the plaster and copied the words of the Law. Consequently, the nations were doomed for having failed to follow the precepts of the Torah. R. Shimeon argues for an alternate scenario, according to which the Law was written upon the plaster, not under it. Moreover, as in the Tosefta, he specifies that the words "That they may not teach you to do according to all [their abominations]," a quotation from Deut 20:18 relating to the Canaanites, was written at the bottom of the stones. This statement is followed by the comment: "Hence you learn that if they turn in penitence, they will be accepted." The Babylonian Talmud thus roughly follows the account of the discussion found in the Tosefta, which includes the quotation of Deut 20:18 and states that the possibility of repenting was offered to the Canaanites.[36] Then we find the conclusion formulated in the Jerusalem Talmud, according to which the nations will burn like

---

35 See the discussion by Saul Lieberman in *Tosefta Ki-Fshutah: Nashim*, 3, 700.
36 The Talmud simply formulates the problem in terms of "under/upon the plaster" rather than "on the stones/on the plaster."

lime (or plaster), an idea that the Babylonian Talmud attributes to R. Yehudah. Finally, the passage goes back to the teaching of R. Shimeon, who quotes yet another biblical verse, "(When you go forth to war against your enemies, and the Lord your God gives them into your hands,) and you take them captive ..." (Deut 21:10) and teaches: "(This is) to include (*lerabot*, lit.: to increase or to gather) the Canaanites who reside in the Land of Israel; so that if they turn in penitence, they will be accepted."[37] As in the Tosefta, Rabbi Shimeon expresses the more lenient view, that it was possible for the Canaanites to escape destruction at the hand of the Israelites, provided the former recognized the authority of the Torah and repented from their evil ways.

4.7 Conclusion

In conclusion, I wish to emphasize that while the texts presented here have a clear exegetical dimension, the main question underlying the admittedly obscure discussions they contain is both theological and ethical. This has to do with the fact that these texts discuss the sharing of religious Law, a particular kind of religious knowledge, and one that has special relevance because of its connection with divine judgement. The basic assumption of the rabbis was that God could punish the nations only if God's Law had first been communicated to the nations. There could be no just punishment if the Law and the punishment of the transgression had not been made known first. Without knowledge, there would be no responsibility. In this case, the possession or lack of knowledge can therefore be said to determine *soteriological* status, and not merely *social* status. The case of the Canaanites represented a particular challenge from a theological and ethical point of view, because according to Scripture their fate was particularly cruel; but the very nature of this challenge was not fundamentally different from the one raised by the confrontation between Israel and the nations in general. The main issue behind the sharing of the Torah remained the same: theodicy.

The problem faced by the rabbis did not simply originate in a kind of conflict between their ethical ideas and the content of Scripture. It was caused by their decision to envision the Torah as a Law given to Israel alone. Indeed, it was

---

37 Here I follow the version of MS Oxford d.20 (2675), consulted on the website of the Academy of the Hebrew Language (Ma'agarim). Other manuscripts add or substitute: "*and you take them captive*, to include the Canaanites who reside outside the Land [of Israel]." This is the version favored in the Soncino translation of the Talmud.

because most of them conceived of the Torah as Israel's exclusive property, while at the same time being aware of the theological and ethical problems this raised, that they had to imagine stories about the nations who refused to live according to the Torah after having heard about its commandments. Thus the question arises: Why did most rabbis break with the idea of the Torah as a universal Law and choose to conceive of the Torah as the particular law of Israel alone? In order to answer this question, we must take into account the historical context in which the Palestinian rabbis lived.

The significance of the rabbinic texts presented here is not merely exegetical and theological; they also illustrate a wider phenomenon in rabbinic literature, i.e. the development of a counter-culture in the context of the Roman Empire where the rabbis who wrote the Mishnah, the Tosefta, the halakic midrashim and the Jerusalem Talmud lived. As Natalie Dohrmann has emphasized, the rabbis were keenly aware of the challenge posed by Roman law and the Roman legal order, and the development of the rabbinic halakic discourse may to a large extent be seen as a response to such a challenge.[38] Therefore the preoccupation of the rabbis with the universal or particular character of the Torah and its theological and ethical consequences for the nations can be properly understood only if placed back into this historical context. In a world dominated by Roman law (even if other legal systems continued to function to a certain extent), most rabbis seem to have chosen the path of cultural resistance by rejecting mimetic rivalry with Roman universalism, even if a certain amount of mimicry can also be identified in the traditions examined above.[39] Most rabbis embraced the notion of the Torah as the exclusive inheritance of Israel, and simultaneously chose to write down their traditions in Hebrew and Aramaic, in an idiosyncratic and elliptical style.[40] In both discourse and practice, most rabbis thus favoured the path of "particularism" and hidden knowledge, which in their historical context may be seen as an expression of cultural resistance.

---

38 Natalie B. Dohrmann, "Law and Imperial Idioms: Rabbinic Legalism in a Roman World," in *Jews, Christians and the Roman Empire: The Poetics of Power in Late Antiquity*, ed. Natalie B. Dohrmann and Annette Yoshiko Reed (Philadelphia: University of Pennsylvania Press, 2012), 63–78; eadem, "Can Law be Oral? The Mixed Message of Rabbinic Oral Law," in *Public and Private in Ancient Mediterranean Law and Religion*, ed. Clifford Ando and Jörg Rüpke (Berlin: De Gruyter, 2015), 187–216.
39 See my forthcoming article, "Rabbinic Universalism Reconsidered." In post-colonial studies, mimicry is described as one of the strategies adopted by subalterns in order to resist imperial or colonial domination. See for instance Leo G. Perdue and Warren Carter, *Israel and Empire: A Post-Colonial History of Israel and Early Judaism* (London: Bloomsbury, 2015).
40 That many Greek and Latin loan-words are found in rabbinic literature does not alter the fact that no rabbinic work is composed in Greek or Latin.

References

Philip S. Alexander, "The Toponymy of the Targumim, with Special Reference to the Table of the Nations and the Boundaries of the Land of Israel" (PhD diss., University of Oxford, 1974).

Clifford Ando, *Imperial Ideology and Provincial Loyalty in the Roman Empire* (Berkeley: University of California Press, 2000).

Victor Aptowitzer, "Les premiers possesseurs de Canaan, légendes apologétiques et exégétiques," *Revue des études juives* 82 (1926): 274–86.

Wilhelm Bacher, "The Supposed Inscription upon 'Joshua the Robber,' Illustrated from Jewish Sources," *Jewish Quarterly Review* 3 (1891): 354–55.

Katell Berthelot, "The Canaanites who 'trusted in God': an original interpretation of the fate of the Canaanites in rabbinic literature," *Journal of Jewish Studies* 62 (2011): 233–61.

Katell Berthelot, "Rabbinic Universalism Reconsidered: The Roman Context of Some Traditions Pertaining to the Revelation of the Torah in Different Languages," forthcoming in *Jewish Quarterly Review*.

Herbert Danby, *The Mishnah* (Oxford: Oxford University Press, 1933).

Natalie B. Dohrmann, "Law and Imperial Idioms: Rabbinic Legalism in a Roman World," in *Jews, Christians and the Roman Empire: The Poetics of Power in Late Antiquity*, ed. Nathalie B. Dohrmann and Annette Yoshiko Reed (Philadelphia: University of Pennsylvania Press, 2012), 63–78.

Natalie B. Dohrmann, "Can Law be Oral? The Mixed Message of Rabbinic Oral Law," in *Public and Private in Ancient Mediterranean Law and Religion*, ed. Clifford Ando and Jörg Rüpke (Berlin: De Gruyter, 2015), 187–216.

Steven Fraade, "Before and After Babel: Linguistic Exceptionalism and Pluralism in Early Rabbinic Literature and Jewish Antiquity," *Diné Israel* 28 (2011): 31*–68*.

Steven Fraade, "The Torah Inscribed/Transcribed in Seventy Languages," in *Hebrew between Jews and Christians*, ed. Daniel Stein Kokin, Studia Judaica 77 (Berlin: De Gruyter, forthcoming).

Shamma Friedman, "The Primacy of Tosefta to Mishnah in Synoptic Parallels," in *Introducing Tosefta: Textual, Intratextual and Intertextual Studies*, ed. Harry Fox and Tirzah Meacham (Hoboken: Ktav, 1999).

Sidney Greidanus, "The Universal Dimension of Law in the Hebrew Scriptures," *Studies in Religion/Sciences Religieuses* 14/1 (1985): 39–51.

Judith Hauptman, *Rereading the Mishnah: A New Approach to Ancient Jewish Texts*, Texts and Studies in Ancient Judaism 109 (Tübingen: Mohr Siebeck, 2005).

Marc (Menahem) Hirshman, *Torah for the Entire World* (Tel Aviv: Hakibbutz Hameuchad, 1999).

Marc (Menahem) Hirshman, "Rabbinic Universalism in the Second and Third Centuries," *Harvard Theological Review* 93 (2000): 101–15.

Marcus Jastrow, *A Dictionary of the Targum, the Talmud Babli and Jerushalmi, and the Midrashic Literature* (New York: Choreb, 1926).

Menahem Kahana, *The Genizah Fragments of the Halakhic Midrashim. Part 1* (Jerusalem: The Hebrew University Magnes Press, 2005).

Menahem Kister, "The Fate of the Canaanites and the Despoliation of the Egyptians. Polemics among Jews, Pagans, Christians, and Gnostics: Motifs and Motives," in *The Gift of the Land and the Fate of the Canaanites in Jewish Thought*, ed. Katell Berthelot, Joseph David, and Marc Hirshman (New York: Oxford University Press, 2014), 66–111.

Jacob Z. Lauterbach, *Mekhilta de-Rabbi Ishmael: A Critical Edition, Based on the Manuscripts and Early Editions, with an English Translation, Introduction, and Notes*, 2nd ed. (Philadelphia: The Jewish Publication Society, 2004).

Carlos Lévy, *Cicero academicus: Recherches sur les "Académiques" et sur la philosophie cicéronienne* (Rome: Ecole française de Rome, 1992).

Hans (Yohanan) Lewy, "Ein Rechtsstreit um Boden Palästinas im Altertum," *Monatschrift für Geschichte und Wissenschaft des Judentums* 77 (1933): 84–99, 172–80.

Saul Lieberman, *Hellenism in Jewish Palestine*, 2nd ed. (New York: The Jewish Theological Seminary, 1994).

Saul Lieberman, *Tosefta Ki-Fshutah: Nashim, 3* (New York: The Jewish Theological Seminary, 1995).

John W. Martens, *One God, One Law: Philo of Alexandria on the Mosaic and Greco-Roman Law* (Boston – Leiden: Brill, 2003).

Hindy Najman, "A Written Copy of the Law of Nature: An Unthinkable Paradox?" *The Studia Philonica Annual* 15 (2003): 54–63.

Leo G. Perdue and Warren Carter, *Israel and Empire: A Post-Colonial History of Israel and Early Judaism* (London: Bloomsbury, 2015).

Fritz F. Von Schwind, *Zur Frage der Publikation im römischen Recht*, Münchener Beiträge zur Papyrusforschung und Antiken Rechtsgeschichte 31 (Munich: Beck, 1940).

Willem Smelik, *Rabbis, Language and Translation in Late Antiquity* (Cambridge: Cambridge University Press, 2013).

Gregory E. Sterling, "Universalizing the Particular: Natural Law in Second Temple Jewish Ethics," *The Studia Philonica Annual* 15 (2003): 64–80.

Herman L. Strack and Günter Stemberger, *Introduction to the Talmud and Midrash* (Minneapolis: Fortress, 1996).

Moshe Weinfeld, "The Ban on the Canaanites in the Biblical Codes," in *History and Traditions of Early Israel*, ed. André Lemaire, Supplements to Vetus Testamentum 50 (Leiden: Brill, 1993), 142–60.

Moshe Weinfeld, *The Promise of the Land: The Inheritance of the Land of Canaan by the Israelites* (Berkeley: University of California Press, 1993), 96–97.

Delfim F. Leão

5 Alexandria, Diaspora, *Politeuma* and *Patrioi Nomoi*: The Sharing and Hiding of Jewish Identity

5.1 Introduction

During the Hellenistic period, most of the former Greek *poleis* continued to exist, at least as urban spaces, although without the autonomy and liberty of movement that they had enjoyed during the Archaic and Classic periods, especially in terms of foreign policy.[1] Because the essence of the Hellenistic state depended on the monarch and on those working more directly with him, the structure of the *polis* ended up being a strange body within this new reality. Even so, it could not simply be eliminated, because of the symbolic importance it had in the past history of Greece. The *poleis* managed thereby to keep the essence of the constitutional apparatus of the past (popular assembly, council, courts, annually elected magistrates), but were now dependent on the will of the king, whose orders had to be obeyed, whether transmitted by letter, by royal regulation (*diagramma*) or by royal ordinance (*prostagma*). Formally, the façade of autonomy was therefore kept, as long as the decrees of the *polis* were moulded according to the instructions of the monarch, which were thus turned into binding laws. Up to a certain point, this situation constituted a fiction tacitly accepted by both parties, because both could extract benefits from it.[2]

Another feature distinctive of the Hellenistic period and of the strategy adopted by Alexander was the founding of new cities, sometimes with a demographic concentration that would have been unthinkable to the classic *poleis*.

---

[1] This work is partially an adapted and expanded version of Delfim F. Leão, "Identity and Cosmopolitanism: The Jewish '*Politeuma*' of Alexandria," in *Alexandrea ad Aegyptum: The Legacy of Multiculturalism in Antiquity*, ed. Rogério Sousa, Maria C. Fialho, Mona Haggag, and Nuno S. Rodrigues (Porto: Afrontamento, 2013), 122–33. I wish to thank Manuel Tröster, who read an earlier version of this paper and whose comments helped me to improve it, especially at the linguistic level. This research was developed under the project UID/ELT/00196/2013, funded by the Portuguese FCT – Foundation for Science and Technology.

[2] At any rate, the payment of a tribute and the acceptance of the presence of royal garrisons, among other charges supported by each individual *polis*, were an unequivocal sign of their dependence on the power of the sovereign.

The most emblematic of those new establishments was certainly Alexandria,[3] a city that would substitute Memphis as the capital of Egypt, under the Ptolemies, the dynasty initiated by the former general of Alexander, who quickly understood how unrealistic it was simply to try to replace the former ruler by another person. Instead of that, he chose to reinforce the stability of the reign of Egypt, an objective that went in accord with the preoccupation of legitimating his power as sovereign.[4]

Identical motivation may explain, at least in part, the construction of the two most emblematic monuments of the new capital: the Museum and the Library. In fact, they both represented, even in antiquity, a vivid illustration of the cosmopolitan spirit of the new Hellenistic cities.[5] Besides that, in the case of the Ptolemies those monuments contributed as well to the purpose of reinforcing the connection with Alexander and of legitimating the authority of a Greek (and hence foreign) matrix in a cultural context as exuberant as that of ancient Egypt.

Despite the importance of those emblematic constructions, the city of Alexandria constituted also a notable ethnic mosaic, where three communities were particularly important: the native Egyptians, the Macedonians and Greeks in general (culturally and politically dominant), and the Jews. Even if it is correct to state that the authority of the pharaoh worked as a coalescing force, fundamental to keep the whole bulk together, there was nevertheless a high risk of disaggregation (or al least of conflict), especially on the part of those who were more ardent in keeping their religious and cultural roots, as happened with the Jews. It is therefore the aim of this paper to discuss, in the next section, the way the cosmopolitanism characteristic of the Hellenistic period (and of Alexandria in particular) managed to deal with the demands of a strong and deeply rooted awareness of Jewish identity.

---

[3] In antiquity, almost twenty cities were founded with the name Alexandria. For a collection of the sources dealing with the cities founded by Alexander, see Waldemar Heckel and John C. Yardley, *Alexander the Great: Historical Sources in Translation* (Malden: Blackwell, 2004), 303–10.

[4] On the strategy adopted by Ptolemy to legitimate his power, see Andrew Erskine, "Culture and Power in Ptolemaic Egypt: The Museum and Library of Alexandria," *Greece and Rome* 42 (1995): 38–48.

[5] Their creation is generally understood as an expression of the Peripatetic influence on this golden period for science, but it also matches a long-lasting tradition of cultural sponsorship, deeply rooted already in the tyrannies of the Archaic and Classical periods, which the new monarchs intended to cultivate as well. See Victor Parker "*Tyrannos*: The Semantics of a Political Concept from Archilochus to Aristotle," *Hermes* 126 (1998): 145–72; Delfim F. Leão, "The *Tyrannos* as a *Sophos* in the *Septem Sapientium Convivium*," in *Symposion and Philanthropia* in *Plutarch*, ed. José Ribeiro Ferreira, Delfim Leão, Manuel Tröster, and Paula Barata Dias (Coimbra: Imprensa da Universidade, 2009), 511–21, at 518–19.

5.2 Greeks and Jews

The trail of contacts between the Greek world and the Jews goes back to a very distant time in the past, as can be inferred from Hebrew names (as Japheth and Javan) reminiscent of Greek mythical names (Iapetos and Ion), and from the fact that king David himself employed, in a period as distant as the tenth century, Greek mercenaries from Crete. On the other side, remains of pottery found in Samaria suggest the existence of commercial contacts with Greece as early as the eighth century. The traditional Athenian emblem of the owl was discovered on Jewish coins minted in the fifth century and, during the Persian invasion, Jewish mercenaries were among the Persian troops that invaded Greece, in 480 BCE, under the orders of Xerxes.[6] One of the earliest significant allusions to the Jews, in Greek literature, occurs in a short reference in the *Histories* of Herodotus (2.104.2–3), concerning the circumcision, a practice that the Syrians of Palestine (i.e. the Jews) adopted from the Egyptians.[7] According to Josephus (*Ag. Ap.* 1.176–182), Clearchus of Soli, a former pupil of Aristotle, related in his first book *On Sleep* that the master had a meeting with a Jew in Asia Minor. The story is usually considered to be apocryphal, but the fact that the Peripatetic Clearchus found the anecdote worthy of record is an indicator of the high opinion held on the Jews (as well as on the Indians) as a people naturally disposed to philosophical reasoning. An approach identically positive is made by Theophrastus, whose testimony (quoted by Porphyry, *Abst.* 2.26) has the undeniable merit of being the earliest source, outside the Bible, to describe the Jewish sacrifices.[8] Among those earliest accounts on Jews made by non-Jews, the largest testimony derives from the work *History of Egypt* written by Hecataeus of Abdera, which is preserved in a long passage quoted by Diodorus of Sicily (*Bibl. Hist.* 40.3). Even if this work contains certain mistakes (as stating that Moses founded Jerusalem and established the sacred temple) and manifests some criticism towards the zealous character of the Jews, as a social characteristic deriving from the harsh experience of exile

---

[6] Cf. Flavius Josephus, *Ag. Ap.* 1.172–173, who derives this information from a Greek poet named Choerilus. See Louis H. Feldman and Meyer Reinhold, *Jewish Life and Thought among Greeks and Romans: Primary Readings* (Edinburgh: T&T Clark, 1996), 1.
[7] For other parallelisms between the Semitic world and Greek literature, from the Homeric poems down to Xenophon, see the detailed systematization of Nuno S. Rodrigues, "Um olhar a Oriente: Imagens do mundo semítico na literatura grega, dos Poemas Homéricos a Xenofonte," in *Génese e consolidação da ideia de Europa, Vol. I: de Homero ao fim da Época Clássica*, ed. Maria do Céu Fialho, Maria de Fátima Silva, and Maria Helena da Rocha Pereira (Coimbra: Imprensa da Universidade, 2005), 335–65.
[8] Even if he also records several mistakes, like stating that sacrifices were made during the night or that humans were used as sacrificial victims.

(40.3.4), Hecataeus presents nonetheless a quite positive image of the Jews, with whom he might have been in direct contact by the time he visited Egypt.

With the reference to Hecataeus of Abdera (who lived c. 360–290 BCE), one reaches a period comprised between the campaigns of Alexander and the beginnings of the dynasty of the Ptolemies, an epoch that shall open a new and gleaming chapter in the history of the Jews, especially in what concerns their establishment in Egypt. Josephus (*Ag. Ap.* 1.186–204) ascribes to this same Hecataeus a treaty *On the Jews*, but its author is, most probably, a Jew that might have composed the work around the middle of the second century.[9] Despite these limitations, one of the passages of Pseudo-Hecataeus quoted by Josephus is quite illustrative of the importance attributed to the respect of traditional regulations among Jews – a feature that Alexander was wise enough to respect, similarly to what he did with other conquered populations, like the Persians. It is therefore pertinent to evoke this episode as an introduction to the question of the privileges that might have been received by the Jews who decided to move to Alexandria[10]:

> Then Hecataeus indicates in turn our attitude toward the laws (*nomoi*), that we choose to suffer anything rather than transgress them, and consider this to be noble. For this reason, he says, though they are verbally abused by their neighbors and by all those who arrive from abroad, as well as being insolently treated on a regular basis by the Persian kings and satraps, they cannot be shifted from their conviction; on the contrary, defenseless they face on behalf of these both tortures and the most terrible of all deaths rather than deny their ancestral ways (*ta patria*). He also provides several evidences of this strong-mindedness in relation to the laws (*nomoi*). He says that when Alexander was on one occasion in Babylon and had decided to clear the temple of Bel which had collapsed, he ordered all his soldiers alike to transport the soil; only the Judeans did not comply, but endured severe beating and paid heavy fines, until the king pardoned them and granted them an amnesty. (Josephus, *Ag. Ap.* 1.190–192)

The presence of Jewish troops serving under Alexander does not constitute a major surprise, because, as discussed above, it was possible as early as the fifth century to find Jewish mercenaries in the Persian army.[11] On the other hand, the

---

[9] For more details on the "discovery" of the Jews by Greek authors, see Feldman and Reinhold, *Jewish Life and Thought among Greeks and Romans*, 1–14, at 10, in what respects the case of Pseudo-Hecataeus analysed here.

[10] The translation is taken from Steve Mason, *Josephus: Against Apion, Translation and Commentary* (Leiden: Brill, 2007), 110–12. The Greek words transcribed in brackets are my addition. The same applies to other passages quoted in translation throughout the paper.

[11] Martin Hengel, "The Interpenetration of Judaism and Hellenism in the pre-Maccabean Period," in *The Cambridge History of Judaism*, ed. William D. Davies and Louis Finkelstein (Cambridge: Cambridge University Press, 1989), 167–228, at 187 and n. 1, says that there is no reason to doubt that Jewish mercenaries served under Alexander, although he considers unhistorical the

idea that the Macedonian leader might have shown indulgence respecting the interdictions dictated by Jewish laws (even facing the risk of some loss of authority)[12] finds a possible parallel in the way he knew to respect former enemies, either because he was convinced that this was the best way of acting or by mere political pragmatism.[13] Thus, besides not being wholly improbable from a historical perspective, this detail is in accord with the tradition that tended to present Alexander as a great benefactor of Jewish identity, to the point of suggesting that this support may have been influenced by divine intervention.

This is the case of the first visit of Alexander to Jerusalem (in 332 BCE), which was preceded by moments of great tension, because the high priest had decided, in a first instance, to remain faithful to Darius, a choice that made the Macedonians march against Jerusalem. The vivid memory of this episode was preserved in Josephus's *Jewish Antiquities* (*Ant.* 11.304–346), in terms whose historicity is, to say the least, highly suspect. Actually, the epiphany of Alexander in Jerusalem has too many points of contact with another experience of divine inspiration – a fact that cannot be ruled out as simple coincidence – lived during the first part of the year 331: the famous pilgrimage of the Macedonian king to the sanctuary of Amon, in the oasis of Siwah (Libya), undertaken in a time when he had already chosen the place where the new capital of Egypt was to be established.[14] Several details adduced when Alexander visits the temple of Jerusalem – like bringing the *Book of Daniel* before him (a book which was in reality written only around 164 BCE), evoking the prophecy that a Greek would overcome the Persian empire – strongly suggest that the episode reflects a later Jewish tradition, in which some usual signs of legendary amplification can be detected in what respects the deeds

---

tradition stating that the Macedonian monarch gave *isopoliteia* to the Judean soldiers that decided to establish themselves in Alexandria. Aryeh Kasher, "The Jewish *Politeuma* in Alexandria: A Pattern of Jewish Communal Life in the Greco-Roman Diaspora," in *Homelands and Diasporas: Greeks, Jews and Their Migrations*, ed. Minna Rozen (London: I.B. Tauris, 2008), 109–25, at 122, sustains on the contrary that the Jewish *politeuma* of Alexandria had "political equality" (*isopoliteia*) enabling its members to "organize independently (of the *polis*) and maintain their own autonomous legal and religious establishments."

12 Mason, *Josephus: Against Apion*, 112 n. 650, comments that the punishment of those disobeying soldiers "seems unnaturally light."

13 On the way, Alexander's behaviour evolved from the image of a leader of a pan-Hellenic colligation against the Barbarians into a strategy of favouring the inclusion of the defeated into the new budding order, see Delfim F. Leão, "Alexandre Magno: da estratégia pan-helénica ao cosmopolitismo," in *Atti del convegno internazionale di studi "Plutarco e l'età ellenistica,"* ed. Angelo Casanova (Florence: Università degli Studi di Firenze, 2005), 23–37.

14 For an analysis of Josephus's report, by comparing aspects of the expedition to Jerusalem with the visit to the sanctuary of Amon, see Joseph Mélèze Modrzejewski, *The Jews of Egypt: From Rameses II to Emperor Hadrian* (Skokie/Ill.: Varda Books, 1995), 50–55.

of the Macedonian leader. Nonetheless, it is still pertinent for the objectives of this analysis to recall the final part of the narrative, where the putative privileges granted by Alexander to the Jews are mentioned[15]:

> And, when the book of Daniel was shown to him, in which he had declared that one of the Greeks would destroy the empire of the Persians, he believed himself to be the one indicated; and in his joy he dismissed the multitude for the time being, but on the following day he summoned them again and told them to ask for any gifts which they might desire. When the high priest asked that they might observe their country's laws (*patrioi nomoi*) and in the seventh year be exempt from tribute, he granted all this. Then they begged that he would permit the Jews in Babylon and Media also to have their own laws (*idioi nomoi*), and he gladly promised to do as they asked. And, when he said to the people that if any wished to join his army while still adhering to the customs of their country (*ethe patria*), he was ready to take them, many eagerly accepted service with him. (Josephus, *Ant.* 11.337–339)

Leaving aside the question of the highly suspect historicity of this report, which moves back to the time of Alexander decisions that were, in fact, taken much later,[16] the essence of the political and ideological meaning of the measures here mentioned may nevertheless be valid. In reality, from a political perspective, this report shows that Judea was able to keep, throughout the Hellenistic period, a position comparable to the one it had during the Persian domination: the capacity to act as an ethnic and religious entity, organized around the priesthood power, whose centre was the sacred temple at Jerusalem. From an ideological viewpoint, the account illustrates the bases for the interrelations that were to be established between the Hellenistic sovereigns and the Jews: the first would distribute benefits and accept to respect the Mosaic law, while the latter would guarantee loyalty to the monarch and the readiness to fight under his command. There was however an important evolution concerning the inner legal nature of the Torah: in the past, it worked for the Jews as a law issued by the central power, binding by itself, but now it was presented as the "ancestral law" (*patrios nomos*) of the Jews, whose validity had to be confirmed by the new rulers. This way the Torah ended up by becoming closer to the juridical statute of the *patrioi nomoi* used by the Greeks of the Asian cities freed by Alexander from Persian rule, thus finding a balanced and ingenious mode of keeping the essence of deeply rooted religious traditions in a new political and social order.

---

15 The translation is taken from Ralph Marcus, *Flavius Josephus: Jewish Antiquities, Books IX-XI* (London: Loeb, 1958), 477–79.
16 Mélèze Modrzejewski, *The Jews of Egypt*, 55, says that Josephus is attributing to Alexander a much later event, thus simply "anticipating by some 130 years the step actually taken by Antiochus III about 200 BCE, when he established the status of Jerusalem in the Seleucid empire."

5.3 A Jewish *Politeuma* in Alexandria?

The above-mentioned possible parallelism between the legal situation of the Jews and that of the Greeks is a question that demands further inquiry, taking as reference the Jewish *politeuma* of Alexandria, whose existence, if historically accepted, would represent an elucidative example of the way the Jews from the Diaspora could organize themselves into stable communities, from a social, political and legal standpoint. Although the existence of this *politeuma* has traditionally been accepted, some scholars actually deny it.[17] In fact, despite its prominence, there are no definite proofs that it really existed, even if that is very likely. In a review of communities organized as *politeumata* – representing a specific kind of association, especially during the Hellenistic period – , Patrick Sänger[18] convincingly argues that the term *politeuma* has several meanings and covers a very wide range of realities, such as defining simply a 'political act' of any kind up to the very

---

[17] For a conspectus of the main lines of the debate, see Kasher, "The Jewish *Politeuma* in Alexandria," 109–12. See also, Constantine Zuckerman, "Hellenistic *politeumata* and the Jews: A Reconsideration," *Scripta Classica Israelica* 8/9 (1985–1988): 171–85; Sylvie Honigman, *The Septuagint and Homeric Scholarship in Alexandria: A Study in the Narrative of The Letter of Aristeas* (London: Routledge, 2003), 98–118; Patrick Sänger, "Die Jurisdiktion der jüdischen Gemeinde von Herakleopolis: Normal- oder Sonderfall im hellenistischen Ägypten?" in *Symposion 2015: Vorträge zur griechischen und hellenistischen Rechtsgeschichte*, ed. Delfim F. Leão and Gerhard Thür (Vienna: Verlag der Österreichischen Akademie der Wissenschaften, 2016), 213–32, at 214–18.
[18] Patrick Sänger, "The *Politeuma* in the Hellenistic World (Third to First Century B.C.): A Form of Organisation to Integrate Minorities," in *Migration und Integration—wissenschaftliche Perspektiven aus Österreich Jb 2*, ed. Julia Dahlvik, Christoph Reinprecht, and Wiebke Sievers (Vienna: University Press, 2013), 51–68. The subject is taken up again by him, in a paper written in German that explores the same basic argument, although extending and concretizing the discussion around the meaning of the term *politeuma*: Patrick Sänger, "Das *politeuma* in der hellenistischen Staatenwelt: Eine Organisationsform zur Systemintegration von Minderheiten," in *Minderheiten und Migration in der griechisch-römischen Welt: politische, rechtliche, religiöse und kulturelle Aspekte*, ed. Patrick Sänger, Studien zur historischen Migrationsforschung 31 (Paderborn: Ferdinand Schöningh, 2016), 25–45. For the questions dealing with the concept of *politeuma* in general, see Walter Ruppel, "*Politeuma*: Bedeutungsgeschichte eines staatsrechtlichen Terminus," *Philologus* 82 (1927): 268–312 and 433–54; Arnaldo Biscardi, "Polis, politeia, politeuma," in *Atti del XVII Congresso Internazionale di Papirologia* (Naples: Centro Internazionale per lo Studio dei Papiri Ercolanesi, 1984), 1201–15; Gert Lüderitz, "What is the *politeuma*?" in *Studies in Early Jewish Epigraphy*, ed. Jan Willem van Henten and Pieter W. van der Horst, Arbeiten zur Geschichte des antiken Judentums und des Urchristentums 21 (Leiden: Brill, 1994), 183–225; Mogens Herman Hansen, "Polis, Politeuma and Politeia: A Note on Arist. *Pol.* 1278b6-14," in *From Political Architecture to Stephanus Byzantius. Sources for the Ancient Greek Polis*, ed. David Whitehead (Stuttgart: Franz Steiner, 1994), 91–8; Delfim F. Leão, "*Politeuma* in Plutarch," *Synthesis* 23 (2016): e007. Released November 2016: http://www.synthesis.fahce.unlp.edu.ar/article/view/SYNe007.

specific and technical designation of ethnically categorized communities with a military background that can be described as semi-autonomous administrative units, as they existed in several towns or districts of Ptolemaic Egypt[19]:

> The word *politeuma* is frequently used in the Greek language, and has a wide spectrum of meanings. It can, for instance, refer to a 'political act' or appear as a term for 'government', 'citizenry' or 'state'. As a technical term *politeuma* can, in the context of a Greek city-state or *polis*, also refer to the political leading class of citizens as a sovereign body with specific rights. Therefore, in an oligarchic constitution the word refers to a section of the citizenry; in a democratic one to the entire citizenry. However, the word, as a technical term, is not just restricted to the political organisation of a classical Greek *polis*, but can also be applied to name a specific and organised group of persons within an urban area. In this context we are dealing, apart from one exception (namely a *politeuma* of soldiers in Alexandria [...]), with minorities whose ethnic designation is pointing to a migrant background. The members of such a *politeuma* were concentrated in a certain district of a town, which was initially foreign to them and where they lived as an ethnic community.

From a legal and constitutional perspective, the most complex and also most interesting use of the term *politeuma* is the one mentioned last, which designates a reality that could be found during the Hellenistic period and that seems to be specific to the strategic political planning of the Ptolemies, as an ingenious way of promoting in the regions under their control migrant groups, probably military in their origin and usually sharing the same ethnic roots, by allowing them to govern themselves as administrative units. In fact, eight ethnic *politeumata* have been identified for this period, all of them in areas controlled by the Ptolemies.[20] Two of them have attracted much attention, both consisting of Jewish groups: those of Herakleopolis and of Berenike.[21] The case of Herakleopolis in Middle Egypt is of capital importance, because a group of twenty papyri (*P.Polit. Iud.*, dated between 144/43 and 133/32 BCE) was found there and made a determinant contribution to the understanding of the administrative function of the institution of the *politeuma*. For they show that the officials who governed the Jewish *politeuma* dealt, on the one hand, with disputes that were internal (and sometimes also external) to the community associated to the *politeuma* and, on the other hand, they also provide a good impression of the range of legal issues

---

19 Sänger, "The *Politeuma* in the Hellenistic World (Third to First Century B.C.)," 52. See also Sänger, "Das *politeuma* in der hellenistischen Staatenwelt," 35–8.
20 This is probably true even for the *politeumata* at Sidon. See Sänger, "The *Politeuma* in the Hellenistic World (Third to First Century B.C.)," 53–7 and 61.
21 Unlike the possible (and probable) Jewish *politeuma* of Alexandria, attested only by the so called "Letter of Aristeas," later referred to in this analysis, the *politeumata* of Herakleopolis and of Berenike are corroborated by independent documentation.

these officials covered. The competences they had in the field of justice are comparable to those of Ptolemaic officials, a feature that seems to indicate that *politeumata* resembled semi-autonomous communities whose internal structure had obtained a public dimension, a transformation that was certainly due to a governmental decision. It is therefore quite significant that the institution of *politeumata* by the Ptolemies allowed them to attract and integrate migrant groups who were useful to their kingdom (especially for the army), incorporating them in the upper levels of the population, by giving them a fixed place in the administration of Ptolemaic Egypt.[22]

It is now time to focus again on the case of Alexandria. According to Pseudo-Hecataeus,[23] not long after the battle of Gaza (312), the group of Jews who came to Egypt following the Macedonian conquest brought with them the Torah. Ezekias, the high priest who accompanied them from Judea, gathered a group of friends, possibly during the Sabbath, and read them the whole text, in Hebrew. Still according to Pseudo-Hecataeus, "he had their settlement (*katoikesis*) and the constitution written (*politeia gegrammene*)."[24] The passage is awkward and ambivalent, because the context does not make clear whether the terms *katoikesis* and *politeia* should be understood as being applied to the past history of the Jews or to the very moment when this group established itself in Alexandria.[25] Independent from the way this passage is interpreted, it remains a fact that the Jewish community felt very soon the need of having a Greek translation of the Torah, due perhaps to the fact that the process of *Hellenization* had been so quick

---

[22] See Zuckerman, "Hellenistic *Politeumata* and the Jews"; Sänger, "The *Politeuma* in the Hellenistic World (Third to First Century B.C.)," 63–6; Sänger, "Das *politeuma* in der hellenistischen Staatenwelt," 41–4.
[23] Quoted by Josephus, *Ag. Ap.* 1.186–189.
[24] Josephus, *Ag. Ap.* 1.189. Translated by Mason, *Josephus: Against Apion*, 110.
[25] Hengel, "The Interpenetration of Judaism and Hellenism in the pre-Maccabean Period," 192–3, is also ambivalent in the way he interprets this *politeia gegrammene*, which he tends to identify with a royal decree allowing the Jews to establish themselves in Alexandria with a special statute of ethnic minority. Harald Hegermann, "The Diaspora in the Hellenistic Age," in *The Cambridge History of Judaism*, ed. William D. Davies and Louis Finkelstein (Cambridge: Cambridge University Press, 1989), 115–67, at 160, states that the passage expressly mentions "a short royal decree, the contents of which would be comparable to the letter from Antiochus III to Zeuxis." However, the suggestion that the text was read from the (Hebrew) original may imply, on the contrary, that it was the Torah and that the *politeia* in question was the constitution of the Judean nation. On the other hand, the idea that Ezekias "had been closely in touch with us" (*synethes hemin genomenos*) may be an indication that the high priest was acquainted with the Greeks and with their habits. On the interpretation of this crooked passage and on its connection to the translation of the Septuagint, see also Mélèze Modrzejewski, *The Jews of Egypt*, 99–104; Mason, *Josephus: Against Apion*, 110 n. 636.

that, a few decades after their establishment in Alexandria, most of the Jews were no longer able to understand Hebrew. The first version of the Torah in Greek is the famous translation by the Septuagint, and this is not the time to discuss thoroughly in what conditions it may have been executed. For the purposes of the present paper it is enough to recall two possible (even if not certain) explanations for the making of the translation: first, the aforementioned hypothesis that it was motivated by the insufficient linguistic proficiency in Hebrew of the Jews attending the Synagogue in Alexandria; second, the tradition that it was the successor of Ptolemy I Soter (therefore Ptolemy II Philadelphus) who, around the year 270 BCE, decided to have the Torah translated into Greek, in order to enrich the collections of the Library.[26] According to the same tradition, Demetrius of Phalerum, a former Athenian statesman, was assigned the role of supervising the task.[27] It is not implausible that both reasons may have played a complementary role, and therefore that a practical need of the Jewish community had met the monarch's desire to improve the capacity of the Library (thus widening the access to a text to which part of his subjects attributed capital importance).

This tradition is, in fact, recorded in a document known as the Letter of Aristeas, supposedly written by a courtier, but whose author is most probably a Jew. According to this testimony, the Jewish community and the king himself were so satisfied with the work of the translators that they decided that it should be considered a paradigmatic text and remain unchanged in the future. For the purposes of this analysis, and despite the great importance of the exegetic questions raised by the Bible of the Septuagint, it is the reaction of the Jews and the way the Jewish community is represented that has a more direct interest. Let us evoke then a paraphrase of the Letter of Aristeas provided by Josephus[28]:

> Now, when the Law (*nomos*) had been transcribed and the work of translation brought to an end in seventy-two days, Demetrius assembled all the Jews at the same place where the laws (*nomoi*) had been rendered, and in the presence of the translators read them aloud. Thereupon the people expressed their approval of the elders who had interpreted the Law (*nomos*), and also praised Demetrius for conceiving the idea through which he had become the originator of great benefits to them, and they urged him as well to give their leaders the

---

26 See Feldman and Reinhold, *Jewish Life and Thought among Greeks and Romans*, 17–22, at 18–19.
27 As is remarked by Mélèze Modrzejewski, *The Jews of Egypt*, 100, this attribution to Demetrius is rather awkward, because he "had been unwise enough to favor the succession of the king's eldest son in preference to Philadelphus, falling into disgrace when Philadelphus was made king."
28 Translation by Ralph Marcus, *Flavius Josephus: Jewish Antiquities, Books XII-XIV* (London: Loeb, 1957), 53–5.

Law (*nomos*) to read; and all of them, including the priest and the eldest of the translators and the chief officers (*proestekotes*) of the community (*politeuma*), requested that, since the translation had been so successfully completed, it should remain as it was and not be altered. (Josephus, *Ant.* 12.107–108)

From a political and legal standpoint, this text provides some precious information. The juridical nature of the Torah is insistently underlined by the terms used to refer to it in Greek (*nomos/nomoi*); on the other side, the Jewish community is given the name *politeuma*. In the above-mentioned passage of Pseudo-Hecataeus on the coming of Ezekias to Alexandria (Josephus, *Ag. Ap.* 1.189), it was the word *katoikesis* that was used, a term that, together with the variant *katoikia*, is the one generally employed to define a colony of outsiders in a particular site.[29] This kind of organization implied some capacity of self-government, but not necessarily the civic rights characteristic of a city.[30] *Politeuma* is a word that may also be used to name generically any urban settlement and its inhabitants, although it classifies more in particular a community of alien settlers (even if not specifically Jews), with privileges up to a certain point comparable to civic rights. Another distinctive aspect that deserves being mentioned is that those ethnic groups are regularly characterized by a strong religious identity. As has been suggested,[31] it seems persuasive that the Jewish *politeuma* of Alexandria was military in origin, and that this circumstance may have granted the members of the garrison a distinct and superior status by comparison to the rest of the Jewish community, which constituted the *plethos* of Alexandria in broad sense.

In order to establish *politeumata* and *katoikiai* it would certainly be necessary to have an official authorization. Maybe the above-mentioned *politeia gegrammene* in the passage of Pseudo-Hecataeus about Ezekias could have corresponded to this foundational document, despite the difficulties concerning the interpretation of this expression. On the other side, even if the tradition of the benefits granted by Alexander to the Jews is certainly magnified and at the very least in part anachronistic, it may nevertheless reflect the essence of the conditions given to the first Jewish settlers of Alexandria[32]: the right of living according to their ancestral laws or customs (*patrioi nomoi, idioi nomoi, ethe patria*), and of applying those same

---

[29] On the terminology used in the sources to refer to those relatively autonomous communities, see Hegermann, "The Diaspora in the Hellenistic Age," 158–61.
[30] Nevertheless, sometimes the *politeumata* could develop into cities. There were other designations to name communities of aliens, like *laos*, *synodos* and *synagoge* (although the latter two are later in time).
[31] By Sandra Gambetti, *The Alexandrian Riots of 38 C.E. and the Persecution of the Jews: A Historical Reconstruction* (Leiden: Brill, 2009), 48–9.
[32] See supra commentary on Josephus, *Ant.* 11.337–339.

traditional laws among the persons who voluntarily consider them as binding rules – as long as they did not enter in conflict with the royal authority. The balance between sharing and hiding the force of a traditional religious identity was therefore essential to the success of this kind of social and political integration.

Even without including among those concessions the right of full citizenship (as happened with the Greek and Macedonian communities), this was undoubtedly an intelligent way of promoting mobility and attracting active populations. It also favoured social peace, because *politeumata* like the one that is believed to have existed in Alexandria had the legal capacity of appointing magistrates and of creating their own grid of courts and schools, where the norms of the Mosaic Law could be applied and taught.[33] This reality is, in fact, clearly underlined by another passage in Josephus[34]:

> In Egypt, for example, territory has been set apart for a Jewish settlement (*katoikia*), and in Alexandria a great part of the city (*polis*) has been allocated to this nation (*ethnos*). And an ethnarch (*ethnarches*) of their own has been installed, who governs the people (*ethnos*) and adjudicates suits (*kriseis*) and supervises contracts (*symbolaia*) and ordinances (*prostagmata*), just as if he were the head (*archon*) of a sovereign state (*politeia autoteles*). (Josephus, *Ant.* 14.117).

Apparently, the governing structure was initially almost monarchic, but maybe it did not last long, because the paraphrase of the Letter of Aristeas, previously discussed, refers to a group of "chief officers (*proestekotes*) of the community (*politeuma*)", and not to a single person who concentrated in himself all the authority. It is also not improbable that the governing structure of the *politeuma* may have suffered the effects of a growing Greek influence, as happened with the language and with some more practical procedures, like those involving for example Jewish litigants and Greek judges.[35] In reality and albeit after having followed a very different path, the Greeks of Alexandria and of other Hellenistic

---

[33] Hegermann, "The Diaspora in the Hellenistic Age," 161, accepts that some Jewish colonists may have acquired, as a personal reward, the status of full citizenship, but he maintains (as most scholars do) that the Jews as a community never obtained that right. In the future, this situation would be the cause of significant tensions with the Greek community, as happened when, by the time of Augustus, it was decided to apply taxes to all non-citizens, thereby reducing as well the rights of the Jewish *politeuma* of Alexandria. On this, see Nuno S. Rodrigues, *Iudaei in Vrbe: Os Judeus em Roma de Pompeio aos Flávios* (Lisbon: Calouste Gulbenkian, 2007), 337; Kasher, "The Jewish *Politeuma* in Alexandria," 117–18.
[34] Translation by Marcus, *Flavius Josephus: Jewish Antiquities, Books XII-XIV*, 509.
[35] This is the situation of a certain Dositheos, a Jew of Egyptian origin, who had sued a Jewish woman; their case was judged by a group of Greek magistrates, in a court of Crocodilopolis. On this case, see Mélèze Modrzejewski, *The Jews of Egypt*, 108–9.

cities had reached a set of regulations understood as "common laws" or "civic laws" (*politikoi nomoi*), which remitted not to an archetypical text (as happened with the Jewish Torah), but to a tradition common to several *poleis*, which formed a juridical structure globally identified with the Greek legal experience. The recognition of the binding validity of those traditional determinations (which fell into the broad concept of *patrioi nomoi*) ended up by being one of the most efficient solutions found by the Ptolemies to attract to Egypt many foreigners and to stimulate mobility without putting at risk social peace and the authority of the monarch. In effect, the several Egyptian, Greek and Jewish *nomoi*, to which legal validity was granted, had to be harmonized with the authority of the monarch, who had the ultimate word in the administration of justice, through his regulations and ordinances. But just as the *politikoi nomoi* provided the Greek community with the juridical framework necessary to the political organization and to the resolution of conflicts, the same could have been achieved through the Torah in what respects the Jewish *politeuma*.

As time went by and as a natural result of this confluence of multiple political traditions, the emergence of a common legal substrate should be expected, comparable in its origins and objectives to the process under way in other domains characteristic of this period. Thereby, just as it happened with the linguistic and cultural *koine*, the Hellenistic age (and especially Alexandria) must have favoured also the development of a legal *koine*, responsible as well for the success of the Ptolemies.[36] Thus, they found an acute way of harmonizing the cosmopolitanism originated by the new political and social reality with the necessity to keep a strong identitarian matrix. And, at the same time, this balanced way of sharing and hiding the boundaries of a religious identity, a space for the affirmation of some degree of individuality, was safeguarded, in a universe marked by the confluence of multiple ethnic, political and religious sensibilities.

References

Arnaldo Biscardi, "*Polis, politeia, politeuma,*" in *Atti del XVII Congresso Internazionale di Papirologia* (Naples: Centro Internazionale per lo Studio dei Papiri Ercolanesi, 1984), 1201–15.

Andrew Erskine, "Culture and Power in Ptolemaic Egypt: The Museum and Library of Alexandria," *Greece and Rome* 42 (1995): 38–48.

Louis H. Feldman and Meyer Reinhold, *Jewish Life and Thought among Greeks and Romans: Primary Readings* (Edinburgh: T&T Clark, 1996).

---

36 On the characteristics of this legal *koine*, see Mélèze Modrzejewski, *The Jews of Egypt*, 107–12.

Sandra Gambetti, *The Alexandrian Riots of 38 C.E. and the Persecution of the Jews: a Historical Reconstruction* (Leiden: Brill, 2009).
Mogens Herman Hansen, "*Polis, Politeuma* and *Politeia*: A Note on Arist. *Pol.* 1278b6-14," in *From Political Architecture to Stephanus Byzantius: Sources for the Ancient Greek Polis*, ed. David Whitehead (Stuttgart: Franz Steiner, 1994), 91–8.
Waldemar Heckel and John C. Yardley, *Alexander the Great: Historical Sources in Translation* (Malden: Blackwell, 2004).
Harald Hegermann, "The Diaspora in the Hellenistic Age," in *The Cambridge History of Judaism*, ed. William D. Davies and Louis Finkelstein (Cambridge: Cambridge University Press, 1989), 115–67.
Martin Hengel, "The Interpenetration of Judaism and Hellenism in the pre-Maccabean Period," in *The Cambridge History of Judaism*, ed. William D. Davies and Louis Finkelstein (Cambridge: Cambridge University Press, 1989).
Sylvie Honigman, *The Septuagint and Homeric Scholarship in Alexandria: A Study in the Narrative of The Letter of Aristeas* (London: Routledge, 2003).
Aryeh Kasher, "The Jewish *Politeuma* in Alexandria: A Pattern of Jewish Communal Life in the Greco-Roman Diaspora," in *Homelands and Diasporas: Greeks, Jews and Their Migrations*, ed. Minna Rozen (London: I.B. Tauris, 2008), 109–25.
Delfim F. Leão, "Alexandre Magno: da estratégia pan-helénica ao cosmopolitismo," in *Atti del convegno internazionale di studi "Plutarco e l'età ellenistica,"* ed. Angelo Casanova (Florence: Università degli Studi di Firenze, 2005), 23–37.
Delfim F. Leão, "The *Tyrannos* as a *Sophos* in the *Septem Sapientium Convivium*", in *Symposion and Philanthropia in Plutarch*, ed. José Ribeiro Ferreira, Delfim Leão, Manuel Tröster, and Paula Barata Dias (Coimbra: Imprensa da Universidade, 2009), 511–21.
Delfim F. Leão, "Identity and Cosmopolitanism: The Jewish '*Politeuma*' of Alexandria," in *Alexandrea ad Aegyptum: The Legacy of Multiculturalism in Antiquity*, ed. Rogério Sousa, Maria C. Fialho, Mona Haggag, and Nuno S. Rodrigues (Porto: Afrontamento, 2013), 122–33.
Delfim F. Leão, "*Politeuma* in Plutarch," *Synthesis 23* (2016): e007. Released November 2016: http://www.synthesis.fahce.unlp.edu.ar/article/view/SYNe007.
Gert Lüderitz, "What is the *politeuma*?" in *Studies in Early Jewish Epigraphy*, ed. Jan Willem van Henten and Pieter W. van der Horst, Arbeiten zur Geschichte des antiken Judentums und des Urchristentums 21 (Leiden: Brill, 1994), 183–225.
Ralph Marcus, *Flavius Josephus: Jewish Antiquities, Books XII–XIV* (London: Loeb, 1957).
Ralph Marcus, *Flavius Josephus. Jewish Antiquities, Books IX–XI* (London: Loeb, 1958).
Steve Mason, *Josephus. Against Apion: Translation and Commentary* (Leiden: Brill, 2007).
Joseph Mélèze Modrzejewski, *The Jews of Egypt. From Rameses II to Emperor Hadrian* (Skokie/Ill.: Varda Books, 1995).
Victor Parker "*Tyrannos*. The Semantics of a Political Concept from Archilochus to Aristotle," *Hermes* 126 (1998): 145–72.
Nuno S. Rodrigues, "Um olhar a Oriente: Imagens do mundo semítico na literatura grega, dos Poemas Homéricos a Xenofonte," in *Génese e consolidação da ideia de Europa. Vol. I: de Homero ao fim da Época Clássica*, ed. Maria do Céu Fialho, Maria de Fátima Silva and Maria Helena da Rocha Pereira (Coimbra: Imprensa da Universidade, 2005), 335–65.
Nuno S. Rodrigues, *Iudaei in Vrbe: Os Judeus em Roma de Pompeio aos Flávios* (Lisbon: Calouste Gulbenkian, 2007).
Walter Ruppel (1927), "*Politeuma*: Bedeutungsgeschichte eines staatsrechtlichen Terminus," *Philologus* 82 (1927): 268–312 and 433–54.

Patrick Sänger, "The *Politeuma* in the Hellenistic World (Third to First Century B.C.): A Form of Organisation to Integrate Minorities," in *Migration und Integration – wissenschaftliche Perspektiven aus* Österreich Jb 2, ed. Julia Dahlvik, Christoph Reinprecht, and Wiebke Sievers (Vienna: University Press, 2013), 51–68.

Patrick Sänger, "Die Jurisdiktion der jüdischen Gemeinde von Herakleopolis: Normal- oder Sonderfall im hellenistischen Ägypten?" in *Symposion 2015: Vorträge zur griechischen und hellenistischen Rechtsgeschichte*, ed. Delfim F. Leão and Gerhard Thür (Vienna: Verlag der Österreichischen Akademie der Wissenschaften, 2016), 213–32.

Patrick Sänger, "Das *politeuma* in der hellenistischen Staatenwelt: Eine Organisationsform zur Systemintegration von Minderheiten," in *Minderheiten und Migration in der griechisch-römischen Welt: politische, rechtliche, religiöse und kulturelle Aspekte*, ed. Patrick Sänger, Studien zur historischen Migrationsforschung 31 (Paderborn: Ferdinand Schöningh, 2016), 25–45.

Constantine Zuckerman, "Hellenistic *politeumata* and the Jews: A Reconsideration," *Scripta Classica Israelica* 8/9 (1985–1988): 171–85.

Lautaro Roig Lanzillotta
6 Ancient Greek Patterns of Knowledge Transmission and their Continuity in Gnostic Esotericism

It is well known that Gnostics were fond of the esoteric nature of their doctrines. Whether implied by the highly complex myths unfolded in numerous Nag Hammadi treatises, or explicitly referred to at the outset of some other texts, the secret allure of their teaching is at the core of the Gnostic attitude.[1] Secrecy is something that author and readership (or teacher and pupil) share; it is something that connects them with one another. In the transmission of this esoteric knowledge secrecy is a code shared by transmitter and receiver: it ensures the former of the quality of the latter; and the latter of both the quality of the knowledge received and his or her own aptitude as an interpreter of the encrypted message.[2]

Secrecy is consequently not only about the content, but also about the social and psychological dynamics this implies[3]: it endows those who share the secret with the certainty of their higher status that distinguishes them from the rest of humanity who is not in the know.[4] It is therefore not surprising that Gnostic

---

[1] In fact esoteric knowledge is at the core of the Gnostic search for Gnosis. As Roelof van den Broek, "Gnosticism I: Gnostic Religion," in *Dictionary of Gnosis and Western Esotericism*, ed. Wouter J. Hanegraaf (Leiden: Brill, 2006), 403–16, at 404, points out already the final document of the Messina Conference in 1966 proposed to reserve the term "Gnosticism" for the "group of systems of the Second Century A.D." and reserve the term "Gnosis" for "knowledge of the divine mysteries reserved for an élite," namely an esoteric kind of knowledge. *Contra* Michael A. Williams, "Secrecy, Revelation, and late Antique Demiurgical Myths," in *Rending the Veil: Concealment and Secrecy in the History of Religions*, ed. Elliot R. Wolfson (New York: Seven Bridges Press, 1999), 31–58.
[2] Tanya M. Luhrmann, "The Magic of Secrecy," *Ethos* 17 (1989): 131–65, at 136–37 following Georg Simmel, "The Secret and the Secret Society," in idem, *The Sociology of Georg Simmel* (transl. K. Wolff; Glencoe: Free, 1950); Sissela Bok, *Secrets: On the Ethics of Concealment and Revelation* (Oxford: Oxford University Press, 1984 [1982]); Stanton K. Tefft, *Secrecy: A Cross-Cultural Perspective* (New York, 1980); and Beryl L. Bellman, *The Language of Secrecy: Symbols and Metaphors in Poro Ritual* (New Brunswick: Rutgers University Press, 1984).
[3] See already Simmel, "The Secret," 331, according to whom secrecy is "a sociological form that stands in neutrality above the functions of its contents." See also Hugh B. Urban, "The Torment of Secrecy: Ethical and Epistemological Problems in the Study of Esoteric Traditions," *History of Religions* 37 (1998): 209–48, at 220.
[4] Urban, "The Torment of Secrecy," 210: "secrecy is a discursive strategy that transforms a given piece of knowledge into a scarce and precious resource, a valuable commodity, the possession of

esotericism was so furiously rejected by proto-orthodox Christian contemporaries.[5] While secrecy creates a feeling of belonging among those who share it, it repels outsiders and precludes potential sharers.[6] As a result it may easily become socially troublesome.[7] When dealing with Gnostic esoteric knowledge, Christian polemicists resort to two well-known reactions of those who are excluded from the secret: either contempt of its content or denunciation of its subversive character.[8]

In the footsteps of ancient polemicists, Gnostics have recently been described, and perhaps rightly so, as being ideologically resistant,[9] countercultural, against the mainstream social, cultural, philosophical or religious values. Admittedly Gnostic use and application of Graeco-roman (standard) values and beliefs was in some cases so radical that it could be described as countercultural.[10] But when emphasizing the concealed character of their doctrines and transmitting their knowledge in an esoteric way, were Gnostics really that subversive? In the following pages, my intention will be to argue that Gnostic esotericism as such is not per se deviant, subversive or countercultural. Against current views that place it on the "fringes" of first-century Judaism, and more specifically in Jewish apocalypticism,[11] the present paper will argue that when protecting their knowledge

---

which in turn bestows status, prestige, or symbolic capital on its owner"; Wouter J. Hanegraaff, "Esotericism Theorized: Major Trends and Approaches to the Study of Esotericism," in *Religion: Secret Religion*, ed. April D. DeConick (San Francisco: MacMillan, 2016), 155–70, at 163.

5 See below for the attacks against Gnostic esoteric attitudes by Irenaeus and other heresiologists.

6 Simmel, "The Secret"; Luhrmann, "The Magic of Secrecy," 137; Urban, "The Torment of Secrecy," 220.

7 Already noted by Simmel, "The Secret," at the beginning of the twentieth century, these characteristics can be still found in the literature of the beginning of the twenty-first. See for example Albert de Jong, "Secrecy in Antiquity," in *Dictionary of Gnosis and Western Esotericism*, ed. Wouter J. Hanegraaff (Leiden: Brill, 2005), 1050–54.

8 Regarding the subversive character of Christian splinter groups and their alleged human sacrifices or incest during their nightly meetings see Lautaro Roig Lanzillotta, "The Early Christians and Human Sacrifice," in *The Strange World of Human Sacrifice*, ed. Jan N. Bremmer (Leuven: Peeters, 2007), 81–102, at 95–98.

9 Karen L. King, *The Secret Revelation of John* (Cambridge: Harvard University Press, 2006), 166–68.

10 April D. DeConick, "The Countercultural Gnostic: Turning the World Upside Down and Inside Out," *Gnosis: Journal of Gnostic Studies* 1 (2016): 7–35.

11 Guy G. Stroumsa, "From Esotericism to Mysticism in Early Christianity," in *Secrecy and Concealment: Studies in the History of Mediterranean and Near Eastern Religions*, ed. Hans G. Kippenberg and Guy G. Stroumsa (Leiden: Brill, 1995), 289–309; idem, *Hidden Wisdom: Esoteric Traditions and the Roots of Christian Mysticism*, 2nd ed. (Leiden: Brill, 2006).

and reserving it for a small group of initiates, Gnostics were in fact continuing a long tradition well established in Graeco-Roman philosophical schools ever since the Pre-Socratics,[12] that was continued by Classical and Hellenistic philosophical schools, and perpetuated in Late Antiquity by numerous mystery cults and clubs or *collegia*.[13]

This study comprises three sections, the first of which provides an approach to Gnostic esotericism in early Christian polemical works that provide the basis for modern scholarly analyses. The second part revisits the position Gnostics occupied in the wider religious continuum of the first centuries CE. The third, finally, focuses on Gnostic esoteric knowledge transfer, presenting it as a continuation of ancient knowledge transmission patterns. I will close with some concluding remarks.

6.1 Gnostic Esotericism in Antiheretical Literature and Modern Scholarly Approaches

The secret nature of Gnostic doctrines was a favourite target of antiheretical attacks. Justin Martyr, when denying the accusations of human sacrifice and licentiousness levelled against Christians like himself, even resorts to the secrecy of their rites to accuse Marcionites of exactly the same aberrations: "And whether they perpetrate those fabulous and shameful deeds – the upsetting of the lamp, and promiscuous intercourse, and eating human flesh – we know not."[14]

This negative evaluation of secrecy is continued by Irenaeus. Following Jesus's word that "nothing hidden shall not be revealed, nor secret that shall not be made known,"[15] Irenaeus affirms his intention "to make known to you and all your companions those doctrines which have been kept in concealment until now" at the outset of his *Against all Heresies*.[16] In so doing, he

---

[12] Jean Pepin, "L'arcane religieux et sa transposition philosophique dans la tradition platonicienne," in *La storia della filosofia come sapere critico* (Milano: Angeli, 1984) 18–35; De Jong, "Secrecy in Antiquity," 1050–52.
[13] Luther H. Martin, "Secrecy in Hellenistic Religious Communities," in *Secrecy and Concealment*, ed. Hans G. Kippenberg and Guy G. Stroumsa (Leiden: Brill, 1995), 101–21.
[14] Justin, *Apol.* 1.26.6–7 at 7; also 2.12.1–2. See on the issue, Roig Lanzillotta, "The Early Christians and Human Sacrifice," 95.
[15] In reference to Matthew 10:26.
[16] Irenaeus, *Adv. haer.* prol. 2. On Irenaeus position regarding Christian esoteric knowledge, see Stroumsa, *Hidden Wisdom*, 35.

equates openness with light and truth and secrecy with darkness and falsity: if Valentinians hide their teaching by transmitting it mysteriously and secretly, there must be something wrong with it. In this attempt to demonize his opponents' views, he goes a step further in his attack when he compares the esoteric transmission of Valentinian lore to a wild beast that surreptitiously approaches to attack us:

> So there is no longer any need for many words to overthrow their doctrine, since it has been made manifest to all. To illustrate, when some wild beast has hidden himself in a forest and from there makes attacks on others and kills them, one who cuts down the forest, and so brings the wild beast itself in sight, does not strive to capture the beast because he sees that it is really a wild beast. For people can see its attacks, guard themselves against them, throw javelins at it from all sides, wound it, and thus kill that destructive wild beast. (Irenaeus, *Adv. haer.* 1.31.4).[17]

By simply lifting the secrecy that covered their rites and teachings, the light of truth will expose the falsity of Valentinianism.

In line with his predecessors, Pseudo-Hippolytus also begins his *Refutation of all Heresies* by emphasizing his will to reveal everything concealed. His preface to the *Refutation* abounds in references to the secrets of heretics and the mysteries of their teaching that are only transmitted to neophytes. Given that the silence and awe of the heretics have been wrongly interpreted by many as a sign of their piety, he intends now to expose their true nature:

> Now it seems expedient, even at the expense of a more protracted investigation, not to shrink from labour; for we shall leave behind us no trifling auxiliary to human life against the recurrence of error, when all are made to behold, in an obvious light, the clandestine rites of these men, and the secret orgies which, retaining under their management, they deliver to the initiated only. (Pseudo-Hippolytus, *Refutatio*, pref. 5)[18]

When dealing with the secrecy surrounding Gnostic knowledge and teaching, Pseudo-Hippolytus follows in the footsteps of both his predecessors. On the one hand, he follows Irenaeus's notion that truth cannot be hidden and transmitted secretly: if there is something secret this cannot be truth but error. On the other, he follows Justin's strategy, but makes explicit what in his model was implicit:

---

17 English translation according to Dominic J. Unger and John J. Dillon, *St. Irenaeus of Lyons: Against the Heresies*, Ancient Christian Writers 55 (New York: Paulist, 1992), 64–65.
18 English translation according to Alexander Roberts and James Donaldson, *The Ante-Nicene Fathers, vol. 5: revised and chronologically arranged, with brief prefaces and occasional notes by A. Cleveland Coxe* (Edinburgh: T&T Clark, 1885).

Refutatio overtly equates the esoteric knowledge transfer of his opponents to "clandestine rites and secret orgies."[19]

Pseudo-Hippolytus consequently affirms that his will to expose the secrets by bringing them into the light is mainly intended to let people turn from error to truth. But there is more, of course: besides helping to deceive, spreading error and covering their furtive orgies, the secret character of their teaching allows them to conceal the true origin of their views, which is not in the sacred scriptures but in Greek philosophy:

> In order, then, as we have already stated, that we may prove them atheists... (and) that their doctrines have derived their origin from the wisdom of the Greeks ... and from would-be mysteries, and the vagaries of astrologers. (Pseudo-Hippolytus, *Refutatio*, pref. 7–8)[20]

This approach by the first three antiheretical writers is symptomatic of what we will find later on with Clement of Alexandria, with Origen, Tertullian, Eusebius and Epiphanius. Even if in some cases they enrich their attacks with new arguments, in general they remain loyal to their predecessors, repeating and developing their views. For space's sake, therefore, suffice it with this brief overview and let me shortly refer to modern scholarly interpretations of Gnostic esoteric knowledge. Until recently the scholarly discussion regarding Gnostic esotericism has notably remained within the borders laid by the heresiological framework. On one side of the spectrum Guy Stroumsa[21] recognizes the pivotal role played by secrecy among Gnostics, but understands it as an anomaly, as a proof of the "Gnostic" retreat from society at large, which was going to disappear when Christianity realized that its universal salvation was "refractory to esoteric doctrines."[22] Michael W. Williams[23] on the other side denies that Gnostic groups were inherently secretive, claiming that there is not much evidence to affirm that "Gnostics" kept their books, doctrinal information, or rites concealed. In fact, it is in his view arguable that "in some instances, the very composition of written versions of these [Demiurgical] myths was intended to make them more public."[24] As affirmed above, however, in my opinion, Gnostic esotericism simply continues well established patterns of knowledge transmission in ancient society.

---

19 See above footnote 14.
20 English translation according to Roberts and Donaldson, *The Ante-Nicene Fathers*, 5.
21 Stroumsa, *Hidden Wisdom*.
22 Stroumsa, *Hidden Wisdom*, 106.
23 Williams, "Secrecy, Revelation, and late Antique Demiurgical Myths," 31–58.
24 Williams, "Secrecy, Revelation, and late Antique Demiurgical Myths," 54.

6.2 Revisiting the Position of Gnostic Christianity in the Wider Religious Continuum of Late Antiquity

In constructing their own identity, proto-orthodox Christians did their best to present Gnostic Christians as deviant, and as we saw in the previous pages one of the major arguments used to this end was the secret nature of Gnostic doctrines. This is not surprising: the secrecy of nightly meetings had already been used by the Roman Senate to prohibit Bacchanalia in 186 BCE; and this accusation formed the backbone of the charges levelled against Christians at the beginning of the second century.[25] By turning it now against their opponents, polemical writers aligned themselves with the official Roman attitude and contrasted esoteric sectarianism with the universalistic and open attitude of the Church.[26]

But were the Gnostic Christians as aberrant as their proto-orthodox opponents claimed? Given that Greek philosophers such as Plotinus (*Enneads* II 9), Porphyry and later Neoplatonists to a certain extent support the Christian accusations, modern scholars repeat again and again ancient precedents presenting Gnostic

---

[25] Secrecy was always considered to allow to obscure crimes, among other human sacrifice and incest: this was the case of the *boukoloi* (Dio Cassius 71.4.1; Achilles Tatius, *Leucippe and Cleitophon* 3.15.4; Lollianos, *Phoinikika*). See Albert Henrichs, "Pagan Ritual and the Alleged Crimes of the Early Christians. A Reconsideration," in *Kyriakon: Festschrift J. Quasten I*, ed. Patrick Granfield and Josef A. Jungmann (Münster: Aschendorff, 1970), 18–35 and idem, *Die Phoinikika des Lollianos: Fragmente eines neuen griechischen Romans* (Bonn: R. Habelt, 1972), 28–37. The accusations were also directed to political enemies, see Plato, *Rep.* 571C–574E; 619BC (tyrant); Aristotle, *EN* 1448b 24 (Phalaris, tyrant of Acragas in the sixth century BCE); Tatian, *Or. ad Graec.* 34.1–35; Plutarch, *Publ.* 4.1 (supporters of the Tarquinii); Diodorus Siculus 22.5.1; Polyaenus 6.7.2; Plutarch, *De sera* 556D; cf. Aelian, *VH* 14.41 (Apollodorus of Cassandreia); Sallust, *Cat.* 22.1–2; Plutarch, *Cic.* 10.4; Dio Cassius 37.30.3; Florus 2.12.4; Tertullian, *Apol.* 9.9; Minucius Felix, *Oct.* 30.5 (Catilina). On the issue, see G. Marasco, "Sacrifici umani e cospirazioni politiche," *Sileno* 7 (1981): 167–78.

[26] See Robert M. Grant, "Charges of 'Immorality' Against Various Religious Groups in Antiquity," in *Studies in Gnosticism and Hellenistic Religions Presented to Gilles Quispel on the Occasion of his 65th Birthday*, ed. Roelof van den Broek and Maarten J. Vermaseren (Leiden: Brill 1981), 161–70. As for the participants in the Bacchanalia of 186 BCE, these were also suspected of obscure crimes (Livy, 39.8–18) and Apion accused the Jews of sacrificing a Greek every year (Josephus, *C. Apion.* 11.52–113). Magic was also widely suspected of involving human sacrifice (Juvenal 6.550–52) and different individuals were believed to sacrifice children in order to achieve various goals (Dio Cassius 73.16: Didius Julianus; *Historia Augusta* 8.1–2 and Dio Cassius 79.1: Elagabalus). See in general Franz J. Dölger, "'Sacramentum infanticidii': Die Schlachtung eines Kindes und der Genuß seines Fleisches und Blutes als vermeintlicher Einweihungsakt im ältesten Christentum," *Antike und Christentum* 4 (1934): 188–228, at 211–17.

Christians as "ideologically resistant." In these approaches Gnostic depreciation of matter is one of the favourite examples, but lack of ethical concern, libertinism or asceticism, misanthropy, nihilism, etc., repeatedly recur in discussions and overviews of the Gnostic worldview of the last two centuries.[27]

In the last two decades, however, we have learnt to take these views *cum grano salis*, since in most cases they appear to be simple caricatures. In the case of anti-heretical literature, we know that Gnostic Christians and proto-orthodox Christians were claiming the same space in the same religious arena, and consequently that attacking others helped to define their own borders. As far as Plotinus and the later Neoplatonists are concerned, we tend to give more credit to their reports, but recent research has begun to show that their testimony might not be that objective either: if Plotinus' theory of emanations emerged during his seminars and was partly due to the interaction with his Gnostic colleagues, it is nothing but natural that he attacked their views.[28]

As Jonathan Z. Smith affirmed two decades ago:

> The issue of difference as a mode of both culturally encoding and decoding, of maintaining and relativizing internal as well as external distinctions, raises ... the observation that, rather than the remote "other" being perceived as problematic and/or dangerous, it is the proximate "other," the near neighbor, who is most troublesome.[29]

However, in my research in recent years I have to come to realize that Gnostic lore seems to accommodate the religious and philosophical views of its cultural and religious environment quite well. When approached from a systematic perspective and compared with contemporary worldviews, Gnostic theology, cosmology, anthropology, ethics and epistemology seem to continue rather than discontinue religious and philosophical thought of the wider cultural context.[30]

---

27 See an overview of these charges in Karen King, *What is Gnosticism* (Cambridge: Cambridge University Press, 2003), 21–50.
28 See for example John D. Turner, "Transgressing Boundaries: Plotinus and the Gnostics," *Gnosis: Journal of Gnostic Studies* 1 (2016): 56–85, at 58, 68, 76, who refers to Zeke Mazur, "Primordial Self-Reversion and the Gnostic Background of Plotinian Procession," unpublished paper presented at the Annual Meeting of the International Society for Neoplatonic Studies, New Orleans, June 2005; idem, "Notes pour Plotin, Traite 33 (II 9) Contre les Gnostiques. Draft 1," in *Plotin: Oeuvres completes*. Tome 7, ed. Jean-Marc Narbonne, Mauricio Pagotto Marsola, Lorenzo Ferroni, Kevin Corrigan, and John D. Turner, Collection des Universités de France-Association Guillaume Bude (Paris: Les Belles Lettres, forthcoming), 74–76.
29 Jonathan Z. Smith, *Differential Equations: On Constructing the 'Other'* (Arizona: Arizona State University, 1992) 13–14, as quoted by King, *What is Gnosticism*, 25.
30 Thus also Williams, "Secrecy, Revelation, and late Antique Demiurgical Myths,"

From a theological perspective, Gnostic dualism, distinguishing a transcendent god from a creator god, simply made explicit the implicit tensions of the idealistic theology that beginning with Plato reached Descartes, and which concerned theodicy in particular.[31] If God is either the Good or something above it, whence evil? True, Aristotle and Epicurus[32] placed God beyond the *flammantia moenia mundi*, beyond "the flaming walls of the world," as Lucretius would phrase it,[33] and affirmed that he was alien to human suffering. Middle Platonists, however, loyal to Plato's letter, followed the *Timaeus* in distinguishing between, on the one hand, a God exclusively concerned with the divine part in humans and, on the other, lesser *creator* gods in charge of their lower physical being.[34] Plutarch's world soul,[35] Plutarch's Typhon, Numenius's second god,[36] Marcion's righteous god or the Gnostic demiurge[37] are all similar and parallel attempts to free, in a Platonic way, the highest god of any contact with evil.[38]

As far as cosmology is concerned, Gnostic cosmological dualism, expressed by means of either a bipartite or tripartite view of the cosmos, reflects the same interests and preoccupations of the cosmological thinking of the first centuries CE. In

---

31 Lautaro Roig Lanzillotta, "La recepción de Platón, *Timaeus* 28C, en Clemente de Alejandría," in *Filiación: Cultura pagana, religión de Israel, orígenes del cristianismo, VI*, ed. Patricio de Navascués, Manuel Crespo and Andrés Sáez (Madrid: Trotta, 2016), 259–80.
32 See Jan Opsomer, "Demiurges in Early Imperial Platonism," in *Gott und die Gotter bei Plutarch Götterbilder – Gottesbilder – Weltbilder*, ed. Rainer Hirsch-Luipold (Berlin: De Gruyter, 2005), 51–99; see also Lautaro Roig Lanzillotta, "Dios como Padre y artífice en *Moralia* de Plutarco," in *Filiación: Cultura pagana, religión de Israel, orígenes del cristianismo, V*, ed. Patricio de Navascués, Manuel Crespo and Andrés Sáez (Madrid: Trotta, 2013), 139–56, at 142–46.
33 Lucretius, *De rerum natura* 1.62–79, at 73.
34 Plato, *Timaeus* 42D–E, see Roelof van den Broek, "The Creation of Adam's Psychic Body in the Apocryphon of John," in *Studies in Gnosticism and Hellenistic Religions Presented to Gilles Quispel on the Occasion of His 65th Birthday*, ed. Roelof van den Broek and Maarten J. Vermaseren (Leiden: Brill, 1997), 38–57, at 44.
35 See Lautaro Roig Lanzillotta, "Plutarch and the Image of the Sleeping and Waking Soul," in *Immagini letterarie e iconografia nelle opere di Plutarco, Madrid, Università di Salerno/Red Temática Europea "Plutarco,"* ed. Stefano Amendola, Giovanna Pace, and Paola V. Cacciatore (Madrid: Ediciones Clásicas, 2017), 209–22.
36 Édouard Des Places, *Numenius: Fragments* (Paris, 1973); Michael Frede, "Numenius," *ANRW* II.36.2 (1987): 1034–75; Matthias Baltes, "Numenios von Apameia und der Platonische Timaios," *Vigiliae Christianae* 29 (1975): 241–70. According to Proclus, *In Tim.* 1.303.27–304.3, however, Numenius distinguished three divinities: the first two were demiurges; the third allegedly was the cosmos. See, however, Opsomer, "Demiurges," 63–64.
37 See Lautaro Roig Lanzillotta, "Achamot, el Alma del mundo valentiniana, y su relación con el Demiurgo (Ireneo, *Adv. haer.* 1.5)," in *Jornadas sobre la Filiación VII*, ed. Patricio Navascués and Andrés Sáez (Madrid: Trotta, forthcoming).
38 Roig Lanzillotta, "Dios como Padre y artífice," 146–54.

fact, most of the cosmological frameworks we can retrieve, for example, from the Nag Hammadi treatises or the *Corpus Hermeticum*, encompass the worldviews available at the time, namely the Eudoxian model of the world adapted by Aristotle's cosmology, which according to Cicero was standard at that time, and in some cases even Ptolemy's cosmology widespread in second century CE.[39] Indeed, all these cosmologies include the seven planetary layers, the orbit of the fixed stars and a transcendent region with or without compartments alongside the sublunary world.

Similarly, from an anthropological perspective, Gnostic anthropological dualism, either expressed by means of a dichotomous or trichotomous view of man, also echoes religious-philosophical discussions of the first centuries CE.[40] At that time, Platonic idealism opposing ideas to matter had even penetrated Stoic materialism – as Seneca, Epictetus or Marcus Aurelius sufficiently witness[41] – and Jewish anthropological monism, as Wisdom or 4 Maccabees also show.[42]

---

39 See Lautaro Roig Lanzillotta, "The Cosmology of the *Ascension of Isaiah*: Analysis and Re-Assessment of the Text's Cosmological Framework," in *The Ascension of Isaiah*, ed. Jan N. Bremmer, Thomas R. Karmann, and Tobias Nicklas (Leuven: Peeters, 2015), 259–88; idem, "The Apocalypse of Paul (NHC V,2): Cosmology, Anthropology, and Ethics," *Gnosis: Journal of Gnostic Studies* 1 (2016): 110–31, at 122–15.
40 Lautaro Roig Lanzillotta, "Spirit, Soul and Body in Nag Hammadi Literature: Distinguishing Anthropological Schemes in Valentinian, Sethian, Hermetic and Thomasine Texts," *Gnosis: Journal of Gnostic Studies* 2 (2017): 15–39.
41 The *Platonisierende Stoiker*, according to Zeller (1903) III2, 256. Recent scholarship, however, attempts to minimize Plato's influence: Seneca, see Brad Inwood, "Seneca, Plato and Platonism: the Case of Letter 65," in *Platonic Stoicism and Stoic Platonism*, ed. Mauro Bonazzi and Christoph Helmig (Leuven: Leuven University Press, 2008), 149–67; John M. Rist, "Are You A Stoic? The Case of Marcus Aurelius," in *Jewish and Christian Self-Definition, 3: Self-Definition in the Graeco-Roman World*, ed. Ben F. Meyer and Ed P. Sanders (London: SCM, 1982), 23–45, at 31, who refers to Marcus Aurelius's bipartite or tripartite anthropological schemes only in passing; and Christopher Gill, "Marcus Aurelius' *Meditations*: How Stoic and How Platonic?," in *Platonic Stoicism, Stoic Platonism*, ed. Mauro Bonazzi and Christoph Helmig (Leuven: Leuven University Press, 2007), 189–206. For early Stoics, see Heinrich von Staden, "Body, Soul, and Nerves: Epicurus, Herophilus, Erasistratus, the Stoics, and Galen," in *Psyche and Soma: Physicians and Metaphysicians on the Mind-Body Problem From Antiquity to Enlightenment*, ed. John Wright and Paul Potter (Oxford: Oxford University Press, 2003), 79–116, at 96–105.
42 See 2 Maccabees 3:16–17; 7:37; 14:38; 15:30; Wisdom of Solomon 8:19–20; see James M. Reese, *Hellenistic Influence on the Book of Wisdom and Its Consequences* (Rome: Biblical Institute Press, 1970). Wisdom of Solomon 2:2–4 describes the process of the death of both soul and body in diverse ways. According to Wisdom of Solomon 9:15, the perishable body clearly burdens the soul in a way comparable to Plato's *Phaedo* (81C 20). However, scholars such as Martin Neher, *Wesen und Wirken der Weisheit in der Sapientia Salomonis* (Berlin: De Gruyter, 2004), 131–33, in the wake of Dieter Georgi, "Weisheit Salomos," in idem, *Jüdische Schriften aus hellenistisch-römischer Zeit, III.4* (Gütersloh: Mohn, 1980), and Otto Kaiser, *Grundriss der Einleitung in die kanonischen und deuterokanonischen Schriften des Alten Testaments, 3: Die poetischen und*

When we look at ethics, we get the exact same picture. Gnostic ethics show the same concerns as all contemporary ethical views that conceived of human life as a contest in which individuals needed to control all external distractions in order to focus on their godly and interior realm. Seneca, Philo, Epictetus, Plutarch, Alcinous, Maximus of Tyre, and Hermetic authors widely echo these views. Most interestingly, Gnostic ethical views reflect the same Platonic-peripatetic basis we find in all Middle Platonic authors, who saw the soul as a battlefield of passions and reason, and distinguished within it a rational part alongside an irrational one.

Last but not least, epistemology also shows close similarities. In line with the idealistic cosmology sketched above, Gnostics conceived of the world we live in as illusory. In combination with their anthropological dualism, this opinion resulted in a rather pessimistic worldview: not only do we live in an illusory reality; we are also unaware of it and have scant possibilities of coming to know it. Due to their physical nature, our sensory means further prolong our ignorance and our captivity. It is only in combination with a strict ethical process of detachment from our physical nature and environment that we can regain true knowledge: however, most philosophical and religious movements of the period claim the need to leave behind a discursive kind of knowledge to attain an intuitive, immediate and supra-rational awareness of the divine.[43]

When approached from this systematic perspective Gnostic Christians seem therefore to be rather consequent with the religious and philosophical discourse of their time. They seem to reflect rather acculturation than cultural resistance, as did in fact their proto-orthodox co-Christians:

a. Against reigning polytheism or theological dualism, proto-orthodox Christians claimed monotheism[44];

---

weisheitlichen Werke (Gütersloh: Gütersloher, 1994), 118–19, 199 consider these ideas later additions to the text. The dualistic view of man can even be found in Philo, who sometimes retains the positive Jewish view of the body: see Eduard Schweizer, "Die hellenistische Komponente im neutestamentlichen sarx-Begriff," *Zeitschrift für die neutestamentliche Wissenschaft* 48 (1957): 237–53, at 246–50.

43 On the issue, see A. Hilary Armstrong, "The Hidden and the Open in Hellenic Thought," *Eranos Jahrbuch* 54 (1985): 81–117, at 99: "This knowledge is indeed hidden from most men, but it is held to be attainable by philosophers, even if with great difficulty and perhaps only for brief periods. It is therefore the more impressive that it was this great intellectual tradition which, by its own processes and in its own way, came in the end to the recognition of a final mystery." See Robert Lamberton, "The *aporrhetos theoria* and the Roles of Secrecy in the History of Platonism," in *Secrecy and Concealment*, ed. Hans G. Kippenberg and Guy G. Stroumsa (Leiden: Brill, 1995), 139–52.

44 As a matter of fact until so not so long ago, monotheism was seen as characteristic of Judaeo-Christian religion: the volume *Pagan Monotheism in Late Antiquity*, ed. Polymnia

b. Proto-orthodox cosmology claims a holistic view of the universe as product of the only loving and caring God – something that was going to create the important problems from the point of view of theodicy, which were to remain in the building of Christian theology.
c. While the surrounding cultural world accepted anthropological dualism at face value, and either distinguished two or three elements in man, proto-orthodoxy stubbornly stuck to anthropological monism, of which Irenaeus provides sufficient testimony.[45]
d. Consistently with this view of things, *metriopatheia*, or the Platonic-peripatetic approach to passions, left room for a more radical attitude that of their complete eradication which, even if abandoned by Stoics altogether, gave ground to the most extreme forms of asceticism and sometimes even auto mutilation.[46]
e. Epistemology finally seemed also to present a resistant attitude to epistemological standards of the period by rejecting all human possibility, even capacity, to know, very well worded by Tertullian's quasi nihilistic assert: *credo quia absurdum*.[47]

Despite ancient proto-orthodox attacks and modern scholarly views, I think we may safely affirm that while Gnostic Christians widely accommodated to the views of their neighbours in the wide religious continuum of the historical period, it is proto-orthodox Christians who consistently and systematically rejected everything from their surrounding cultural context.

Athanassiadi and Michael Frede (Oxford: Oxford University Press, 1999) and the following studies included in it: Martin L. West, "Towards Monotheism," 21–40; Michael Frede, "Monotheism and Pagan Philosophy," 41–68; see also Frederick E. Brenk, "Plutarch's Middle-Platonic God: About to Enter (or Remake) the Academy," in *Gott und die Götter bei Plutarch: Götterbilder, Gottesbilder, Weltbilder*, ed. Rainer Hirsch-Luipold (Berlin: Mohr Siebeck, 2005), 27–50; David Sedley, "The Origins of Stoic God," in *Traditions of Theology: Studies in Hellenistic Theology, its Background and Aftermath*, ed. Dorothea Frede and André Laks (Leiden: Brill, 2002), 41–84; and more recently Jan Opsomer, "Plutarch on the One and the Dyad," in *Greek and Roman Philosophy 100 BC–200 AD*, ed. Robert W. Sharples and Richard Sorabji (London: Institute of Classical Studies, University of London, 2007), 379–95.

45 See, for example, Irenaeus's desperate efforts to read 1 Thessalonians 5:23 in a monist way Irenaeus, in *Adv. haer.* 5.6.1.

46 See the story (legendary or not) transmitted by Eusebius (*Historia Ecclesiastica* VI.8) regarding Origen's self-castration, on which see John Anthony McGuckin, *The Westminster Handbook to Origen* (Louisville, Kentucky: Westminster John Knox Press, 2004), 6, 14.

47 Tertullian, *De carne Christi* V, 4. See P. Bühler, "Tertullian: The Teacher of the *credo quia absurdum*," in *Kierkegaard and the Patristic and Medieval Traditions*, ed. Jon B. Stewart (Oxford: Routledge, 2008), 131–44.

But what can we say about the theme of this book? When transferring their knowledge in a concealed way, when creating obstacles in the form and in the content for the dissemination of their esoteric *gnosis* or knowledge, were Gnostics deviant or subversive?

6.3 Gnostic Esoteric Knowledge Transfer: Cultural Continuity or Discontinuity?

The Nag Hammadi treatises widely support the Church Fathers's description of the esoteric transmission of knowledge among Gnostic Christians. Gnostics claimed to have gnosis, which is the Greek word for "knowledge," a knowledge the possession of which liberated those who possessed it from the constrictions of the physical world. Such knowledge was obviously not of a normal human kind, easy to get and available to everyone by means of human reason. Rather, it was divine knowledge disclosing the divine origins of man and explaining how and why he had been degraded to this remote region of the universe.

The *Gospel of Thomas* affirms from the outset the redeeming power of a good interpretation of Jesus's hidden words.[48] Conceived of as a scarce commodity, sometimes they protected their doctrines with a riddle-like language, and other times with complex mythological structures. An example of the former is Saying 50 of the *Gospel of Thomas*, where Jesus says:

> If they say to you, "Where did you come from?," say to them, "We came from the light, the place where the light came into being on its own accord and established itself and became manifest through their image." If they say to you, "Is it you?," say, "We are its children, we are the elect of the living father." If they ask you, "What is the sign of your father in you"?, say to them, "It is movement and repose." (*GosThom* [NHC II,2] 50)[49]

An example of the latter is the description of the unfolding of the divine realm, celestial region and sublunary world in the *Secret Book of John*, the beginning of which already announces the secret character of the instruction included in the book: "The teaching of the saviour and the revelation of the mysteries and

---

[48] *GosThom* (NHC II,2), prol.
[49] English Translation according to Thomas Lambdin, "Gospel of Thomas," in *The Nag Hammadi Codex II,2–7, Together with XIII,2*, Brit. Lib. Or.4926(1), and P. Oxy. 1, 654, 655 with Contributions by Many Scholars*, ed. Bentley Layton (Leiden: Brill, 1989), 53–93.

the things hidden in silence."[50] Also the end of the text emphasizes the esoteric character of the *Apocryphon*, when the Saviour says to John: "I am saying these things to you that you might write them down and give them secretly to your fellow spirits."[51] The teachings are intended to be written down and given "... secretly to your fellow spirits, for this mystery is that of the immovable race."[52]

The secrecy surrounding Gnostic doctrines was of course socially and psychologically productive, since it allowed them to establish both their identity and their social value.[53] *The Gospel of Truth*, for example, presents itself as the gospel that "was revealed to those who are perfect through the mercies of the Father, the hidden mystery."[54] Having the interpretive key could allow interpreters to access a new realm of meaning. *GosTruth* even suggests a salvific meaning in the crucifixion by means of its use of biblical images and expressions: Christ was "nailed to a tree and became a fruit of the knowledge of the Father."

Most importantly, however, possessing the hermeneutical clue to unravelling the *arcana mundi et dei* could free them from the oppression of daily physical existence.[55] After affirming the esoteric nature of the knowledge enclosed in the *logia*, the prologue to the *Gospel of Thomas* promises its interpreters eternal life: "Whoever finds the interpretation of these sayings will not experience death."[56] The individual and his experience, therefore, are also at the core of these secret doctrines. These teachings not only included the passwords for the soul's journey through the celestial realm as Clement of Alexandria claimed,[57] but encouraged introspection, by means of which the individual could transform his self and his life:

> Let him who seeks continue seeking until he finds. When he finds, he will become troubled. When he becomes troubled, he will be astonished, and he will rule over the All. (*GosThom* (NHC II,2] 2).

---

[50] *ApJohn* (NHC II,1) 1–4. English translation according to Michael Waldstein and Frederik Wisse, *The Apocryphon of John: Synopsis of Nag Hammadi codices II,1, III,1, and IV,1 with BG 8502,2* (Leiden: Brill, 2000).
[51] *ApJohn* (NHC II,1) 31.29–31.
[52] *ApJohn* (BG) 75.15–20.
[53] Simmel, "The Secret"; Luhrman, "The Magic of Secrecy," 137; Urban, "The Torment of Secrecy," 220.
[54] *GosTruth* (NHC I,3) 18.11–15.
[55] De Jong, "Secrecy in Antiquity," 1051.
[56] *GosThom* (NHC II,2), prol.
[57] Clement of Alexandria, *Exc. Theod.* 78.

The discovery of essential truths beneath daily experience may perhaps astonish searchers, but their holistic or overarching value will help them to supersede the fragmented nature of their existence. It is the discovery of this unity that allows individuals both to transform themselves and their lives.

We can see that in dealing with the transmission of this knowledge, Gnostic writings consequently fit the three common denominators of every secret tradition in Antiquity very well,[58] since they claim to provide:

a. Access to true reality, to essential truths about the cosmos, gods or origins which are normally concealed under the surface of daily experience
b. A holistic explanation of reality in order to overcome a fragmented reality – the world, the divine, the human being, the soul or mind are in fact interrelated
c. Certain power that transforms the person's self and life.

These three aspects also characterized Greek philosophy from its very beginnings.

As I have shown elsewhere, knowledge dissemination in ancient Greece always encountered both natural and cultural obstacles.[59] Practical knowledge about crafts, trade and religious duties was traditionally protected by the closed circle of the guild structure or family within which it was transmitted.[60] In this context, the master-pupil or father-son relationship and orality in the transmission of knowledge naturally excluded the participation of outsiders. The *Hippocratic Oath*, for example, uses the family image: when accepted into the family-based structure, the pupil is "adopted" and from then onwards behaves as a member of the family.[61]

The highly competitive nature of Greek society was a major cultural impediment. In the Archaic Period, knowledge was increasingly regarded as cultural capital and, consequently, protected as a scarce commodity. Greek intellectuals of the period reflect the same animosity that Hesiod attributes to people of the

---

[58] De Jong, "Secrecy in Antiquity," 1050–54.
[59] Lautaro Roig Lanzillotta, "Concealed Knowledge: Ways of Producing, Protecting, and Sharing Knowledge in Antiquity as a Context for Gnostic Esotericism," 4–11 (forthcoming).
[60] Burkhard Meissner, "Mündliche Vermittlung und schriftliche Unterweisung in der antiken Berufsausbildung," in *Antike Fachschriftsteller: Literarischer Diskurs und sozialer Kontext*, ed. Marietta Horster and Christiane Reitz; Palingenesia 80 (Wiesbaden: Franz Steiner, 2003), 153–75, at 159.
[61] *Hippocratic Oath* 5–6, where the pupil agrees "to hold him who has taught me this art as equal to my parents and to live my life in partnership with him … and to regard his offspring as equal to my brothers in male lineage and to teach them this art." On the combination of written and oral transmission of Hippocratic lore, see Lawrence M. V. Totelin, *Hippocratic Recipes: Oral and Written Transmission of Pharmacological Knowledge in Fifth- and Fourth-Century Greece* (Leiden: Brill, 2009), 22–66 ("Oral Transmission of Medical Knowledge and Written Recipes"), at 22.

same guild[62]: Solon (fr. 20 West) criticized his fellow poet Mimnermus; Heraclitus attacked Homer, Hesiod, Archilochus, Pythagoras, Xenophanes and Hecataeus; Xenophanes did the same to Homer, Hesiod, Pythagoras and Simonides; and Simonides censured the sage Pittacus (fr. 542.ll-6 Page).[63]

Concealing and revealing for the first time clearly develop as forms of social interaction; knowledge is conceived of as a scarce commodity and is consequently the door to social promotion. All Presocratics claim to have the key to understanding reality in a better way than their predecessors, and they also protect their knowledge in a variety of ways. Not only Pythagoras and the Orphic wandering priests protected their doctrine with the institution of an esoteric group around it; Heraclitus, Empedocles and Parmenides also presented, protected and claimed knowledge as a personal achievement. On the one hand,[64] they affirm that they provide access to living truths underlying ordinary things[65]; on the other, their teaching reveals the holistic truths hidden behind a fragmented reality[66]; most importantly, their knowledge is aimed at transforming the self and the lives of their disciples.[67]

---

62 Hesiod, *Works and Days* 25–26, "the potter is against the potter, beggar against beggar, and singer against singer."
63 Jan N. Bremmer, "Religious Secrets and Secrecy in Classical Greece," in *Secrecy and Concealment*, ed. Hans G. Kippenberg and Guy G. Stroumsa (Leiden: Brill, 1995), 61–78, at 69–70.
64 De Jong, "Secrecy in Antiquity," 1051–52.
65 According to Heraclitus the true reality of things is hidden: Not only do we find that "Nature [physis] likes to conceal itself" (B 123 DK) but also "an an-apparent connection is stronger than an apparent one" (B 54 DK). However, the many tend to believe otherwise (B 17 DK): "The many do not take heed of such things as those they meet with, nor do they recognize them when they are taught, though they think they do" (οὐ γὰρ φρονέουσι τοιαῦτα πολλοί, ὁκόσοι ἐγκυρεῦσιν, οὐδὲ μαθόντες γινώσκουσιν, ἑωυτοῖσι δὲ δοκέουσι). Parmenides also claims his Way of Truth accesses the true being underneath the Seeming Way based on what sensorial perception conveys to the majority (B 2 DK). See Empedocles's attack on common opinion in order to claim that truth lies beneath appearances, a true reality for which he can provide the key (B 11 DK).
66 Wisdom is for Heraclitus "...to know the thought by which all things are steered through all things (B 41 DK: ἓν τὸ σοφόν, ἐπίστασθαι γνώμην, ὁτέη ἐκυβέρνησε πάντα διὰ πάντων)"; for Empedocles, Love and Strife give coherence to the manifold accidents due to the combination of the four elements (B 17 DK); and the same might be said of Parmenides, for whom Being, in spite of appearances, is indivisible and homogeneous (B 4 DK), and who asserts (B 5 DK) "it is all one to me where I begin; for I shall come there again in time."
67 The Pythagorean way of life, dietary prescriptions and the shared secret are intended to transform the neophyte. For Heraclitus it is the consciousness of the "common logos" that enables this transformation, which is achieved by introspection, by his self-search as expounded in his "I searched for myself" (B 101 DK). As far as Empedocles is concerned, Peter Kingsley, *Ancient Philosophy, Mystery, and Magic: Empedocles and Pythagorean Tradition* (Oxford: Oxford University Press, 1995), has called attention to the mystery terminology and

In this trend, the Hippocratic *nomos* also endowed medical lore with a religious allure and protected knowledge from outsiders, only transmitting it to the closed group of the initiated: "Things that are holy are revealed only to men who are holy. The profane may not learn them until they have been initiated into the mysteries of knowledge."[68] All philosophical schools, from the Academy[69] to the Stoa,[70] not only continue this esoteric view of knowledge and its transmission,

---

seeds-words metaphor we find in his *Peri Physeos*: Words are seeds that when properly cared for and grown enact the inner transformation of the individual.

68 See the Hippocratic nomos, *Law* 5: Τὰ δὲ ἱερὰ ἐόντα πρήγματα ἱεροῖσιν ἀνθρώποισι δείκνυται· βεβήλοισι δὲ, οὐ θέμις, πρὶν ἢ τελεσθῶσιν ὀργίοισιν ἐπιστήμης. See Emile Littré, *Oevres complètes d'Hippocrate* [1844] IV 642. On the issue, see Walter Burkert, "Craft versus Sect: The Problem of Pythgoreans and Orphics," in *Jewish and Christian Self-Definition, III: Self-Definition in the Graeco-Roman World*, ed. Ben F. Meyer and Ed P. Sanders (London: SCM, 1982), 1–22, at 8; and, in the same line, Vivian Nutton, "Healers in the Medical Market Place: Towards a Social History of Graeco-Roman Medicine," in *Medicine in Society: Historical Essays*, ed. Andrew Wear (Cambridge: Cambridge University Press, 1992), 15–58 n. 17.

69 For Plato's philosophy, see *Symp.* 209E–210A, Socrates compares philosophical inquiry with the initiation into the mysteries. For the culmination of the philosopher's enterprise as *epopteia* or "revelation, vision," see (besides *Symp.* 210A) *Epistle* VII 344B; *Phaedr.* 250A–251A; *Phaed.* 69C–D. Not only Édouard Des Places, "Platon et la langue des mystères," in *Études platoniciennes*, 1929–1979, Études préliminaires aux Religions orientales dans l'Empire romain, 90 (Leiden: Brill, 1981), 83–98; but especially Christoph Riedweg, *Mysterienterminologie bei Platon, Philon und Klemens von Alexandrien* (Berlin: De Gruyter, 1987), have sufficiently shown the extent to which Plato's dialogues, notably the *Symposium* and *Phaedrus*, make use of mystery terminology. See also Heinrich Dörrie, "Philosophie und Mysterium," in *Verbum et Signum II: Beitrage zur mediavistischen Bedeutungsforschung Studien zu Semantik und Sinntradition im Mittelalter*, ed. Hans Fromm, Wolfgang Harms, and Uwe Ruberg (Munich: Fink, 1975), 9–24; idem, "Mysterien (in Kult und Religion) und Philosophie," in *Die Orientalischen Religionen im Römerreich*, ed. Maarten J. Vermasseren (Leiden: Brill, 1981), 341–62.

70 Already Cleanthes seems to have held a similar conception of philosophy and described it in a similar esoteric fashion. According to SVF I 538, Cleanthes described the gods as "mystical shapes," the sun as the "torchbearer," the cosmos as the "mysterion" and those who were "filled with the divine" were the "initiated." According to Plutarch, also Chrysippus conceived of theology as an initiation into the secret rites. In *De stoicorum repugnantiis* 1035A–B [= SVF II, 42]) he reports that the Stoic philosopher viewed theology as the last step in the philosophical curriculum and called it τελετή. See also SVF II, 1008, preserved by the *Etymologicum magnum*, which affirms "that the discourses about the things divine are appropriately called initiations." See on the issue Pierre Boyancé, *Etudes sur le Songe de Scipion* (Paris: E. de Boccard, 1936), 116ff; and idem, "Sur les mystères d'Eleusis," *Revue des Études Grecques* 75 (1962): 460–82, at 466. See also Daniel Babut, *La religion des philosophes grecs* (Paris: Presses Universitaires de France, 1974), 172; Jaap Mansfeld, "Providence and the Destruction of the Universe in Early Stoic Thought," in *Studies in Hellenistic Religions*, ed. Maarten J. Vermaseren (Leiden: Brill, 1979), 129–88, at 134–36; Kempe Algra, "Stoic Theology," in *The Cambridge Companion to the Stoics*, ed. Brad Inwood (Cambridge: Cambridge University Press, 2003), 153–78, at 154.

but also follow Plato in conceiving of philosophy as a mystery-like initiation. It is therefore not surprising that both in Antiquity and in modern scholarship, Plato has been credited with having transmitted an esoteric teaching alongside his exoteric writings. Without going now into the scholarly debate regarding Plato's *Ungeschriebene Lehre* or "unwritten doctrine,"[71] his reticence towards writing as a matter of fact places us again in the context of the master-pupil oral transmission.[72]

6.4 Concluding Remarks

In the first centuries CE, mystery cults were widely embraced by Graeco-Roman society. The flourishing of new mystery cults, secret societies and groups properly reflect the new winds that began to blow during Hellenism. So much so that this even affected philosophical enquiry, which from now on was also introduced to the mystery cult pattern: even if mainly rational in character, philosophy is said to provide the first necessary initiation into a mystery-like experience, the last steps of which were supra-rational in character. Not only Neopythagoreans like Theon of Smyrna, but also other Middle Platonists such as Philo, Alcinous or Plutarch, and later on even Clement of Alexandria, claim that the culmination of the initiation into the philosophical mysteries was an *epoptic* or visionary experience.

---

[71] In favour: Konrad Gaiser, *Protreptik und Paränese bei Platon* (Stuttgart: Kohlhammer, 1959); *Platons ungeschriebene Lehre* (Stuttgart: Klett, 1963); Hans J. Krämer, *Arete bei Platon und Aristoteles* (Heidelberg: C. Winter, 1959); and idem, *Der Ursprung der Geistmetaphysik, Untersuchungen zur Geschichte des Platonismus zwischen Platon und Plotin* (Amsterdam: Grüner, 1964); and more recently Giovanni Reale, *Toward a New Interpretation of Plato*, translated from the tenth edition and edited by John R. Catan and Richard Davies (Washington, DC: The Catholic University of America Press, 1997). *Contra*: Eugène N. Tigerstedt, *The Decline and Fall of the Neoplatonic Interpretation of Homer* (Helsinki: Societas Scientiarum Fennica, 1974), Chapter 6 and *Interpreting Plato* (Stockholm: Almqvist & Wiksell International, 1977); William K.C. Guthrie, *A History of Greek Philosophy, 5: The Later Plato and the Academy* (Cambridge: Cambridge University Press, 1978), 418–42.

[72] Sections such as *Phaedrus* (274B–276C), *Theaetetus* 152C, the *Second Letter* (314), and the *Seventh Letter* (340B–345C) clearly express a preference for oral transmission over writing in the context of philosophical instruction. The *Laws* also considered that some doctrines may be "not communicable before the right time (ἀπόρρητον)," since communicating them prematurely may hinder proper understanding (968E 2–5). See Thomas A. Szlezák, "Plato, section E, Criticism of writing," in *Brill's New Pauly: Antiquity volumes*, ed. Hubert Cancik and Helmuth Schneider, Brill Online, 2014, checked 8 February 2014.

In this context it is not surprising that Gnostic Christian groups presented both their teaching and their transmission of knowledge esoterically, something that could only be fully understood after proper initiation into their mysteries. This attitude was not a development of Jewish apocalypticism and the secrecy surrounding its religious experience. Rather, it was no doubt a consequence of Plato's epistemological dualism, which divided human knowledge into knowledge of the visible world and knowledge of the higher realities behind it, those principles upon which it depends. In this sense, it required the preliminary use of discursive means to break through the delusion of externals in order to allow in a last stage the vision of the divine. Even if accessible after much effort and for a short while, this knowledge was granted now to the Gnostic as it used to be granted to the philosopher.[73]

As I have already affirmed, Gnostic groups had no problem accommodating their Christian views to the Graeco-Roman setting in which they lived. Even if they attitude may have been subversive or deviant, there was nothing culturally resistant in the esoteric nature of their doctrines. In this they were following the pattern provided by all philosophical schools of the period, associations, and clubs or *collegia*. I therefore even wonder whether we can use the term "accommodate" in their case, since strictly speaking they were not accommodating themselves to a different culture; they were simply expressing their views and beliefs according to the cultural standards of the society they were part of, to the environment in which they lived.

References

Keimpe Algra, "Stoic Theology," in *The Cambridge Companion to the Stoics*, ed. Brad Inwood (Cambridge: Cambridge University Press, 2003), 153–78.
Arthur H. Armstrong, "The Hidden and the Open in Hellenic Thought," *Eranos Jahrbuch* 54 (1987): 81–117.
Polymnia Athanassiadi, and Michael Frede, eds., *Pagan Monotheism in Late Antiquity* (Oxford: Oxford University Press, 1999).
Daniel Babut, *La religion des philosophes grecs* (Paris: Presses Universitaires de France, 1974).
Matthias Baltes,"Numenios von Apameia und der Platonische Timaios," *Vigiliae Christianae* 29 (1975): 241–70.
Beryl Larry Bellman, *The Language of Secrecy: Symbols and Metaphors in Poro Ritual* (New Brunswick, NJ: Rutgers University Press, 1984).

---

[73] See above in footnote 43 the text by Armstrong, "The Hidden and the Open in Hellenic Thought."

Hans-Dieter Betz, "Secrecy in the Greek Magical Papyri," in *Secrecy and Concealment: Studies in the History of Mediterranean and Near Eastern Religions*, ed. Hans Kippenberg and Guy Stroumsa (Leiden: Brill, 1995), 153–75.
Georges Boas, "Ancient Testimony to Secret Doctrines," *The Philosophical Review* 62 (1953): 79–92.
Sissela Bok, *Secrets: on the Ethics of Concealment and Revelation* (Oxford: Oxford University Press, 1984) [New York: Pantheon, 1982].
Kees W. Bolle, ed., *Secrecy in Religions,* Studies in the History of Religions 49 (Leiden: Brill, 1987).
Pierre Boyance, "Sur les mystères d'Eleusis," *Revue des Études Grecques* 75 (1962): 460–82.
Pierre Boyance, *Etudes sur le Songe de Scipion* (Paris: E. de Boccard, 1936).
Jan N. Bremmer, "Religious Secrets and Secrecy in Classical Greece," in *Secrecy and Concealment: Studies in the History of Mediterranean and Near Eastern Religions*, ed. Hans Kippenberg and Guy Stroumsa (Leiden: Brill, 1995), 61–78.
Frederick E. Brenk, "Plutarch's Middle-Platonic God: About to Enter (or Remake) the Academy," in *Gott und die Götter bei Plutarch: Götterbilder, Gottesbilder, Weltbilder*, ed. Rainer Hirsch-Luipold (Berlin: Mohr Siebeck, 2005), 27–50.
Luc Brisson, "Premises, Consequences, and Legacy of an Esotericist Interpretation of Plato," *Ancient Philosophy* 15 (1995): 117–34.
Roelof van den Broek, "Gnosticism and Hermeticism in Antiquity: Two Roads to Salvation," in *Gnosis and Hermeticism from Antiquity to Modern Times*, ed. Roelof van den Broek and Wouter J. Hanegraaff (Albany: State University of New York Press, 1988), 1–20.
Roelof van den Broek, "Religious Practices in the Hermetic Lodge: New Light from Nag Hammadi," in *From Poimandres to Jacob Bohme: Gnosis, Hermetism and the Christian Tradition*, ed. Roelof van den Broek and Cis van Heertum (Amsterdam: In de Pelikaan, 2000), 78–95.
Roelof van den Broek, "The Creation of Adam's Psychic Body in the Apocryphon of John," in *Studies in Gnosticism and Hellenistic Religions Presented to Gilles Quispel on the Occasion of His 65th Birthday (Education and Society in the Middle Ages and Renaissance)*, ed. Roelof van den Broek and Maarten J. Vermaseren (Leiden: Brill, 1997), 38–57.
Roelof van den Broek, "Gnosticism I: Gnostic Religion," in *Dictionary of Gnosis and Western Esotericism*, ed. W. Hanegraaf et al. (Leiden / Boston, 2006), 403–16.
Roelof van den Broek and Wouter J. Hanegraaff, eds., *Gnosis and Hermeticism from Antiquity to Modern Times* (Albany: State University of New York Press, 1988).
Pierre Bühler, "Tertullian: The Teacher of the *credo quia absurdum*," in *Kierkegaard and the Patristic and Medieval Traditions*, ed. Jon Bartley Stewart (Oxford: Routledge, 2008), 131–44.
Walter Burkert, "Craft versus Sect: The Problem of Pythgoreans and Orphics," in *Jewish and Christian Self-Definition*, III: *Self-Definition in the Graeco-Roman World*, ed. Ben F. Meyer and Ed P. Sanders (London: SCM, 1982), 1–22.
Walter Burkert, "Das Proömium des Parmenides und die Katabasis des Pythagoras," *Phronesis* 14 (1969): 1–30.
Walter Burkert, "Der geheime Reiz des Verborgenen: Antiken Mysterienkulte," in *Secrecy and Concealment: Studies in the History of Mediterranean and Near Eastern Religions*, ed. Hans Kippenberg and Guy Stroumsa (Leiden: Brill, 1995), 79–100.
Walter Burkert, *Ancient Mystery Cults* (Cambridge: Harvard University Press, 1987).
Walter Burkert, *Lore and Science in Ancient Pythagoreanism* (Cambridge, MA: Harvard University Press, 1972).

Brian P. Copenhaver, *Hermetica: the Greek Corpus Hermeticum and the Latin Asclepius in a New English Translation* (Cambridge: Cambridge University Press, 1992).
April D. DeConick, "The Countercultural Gnostic: Turning the World Upside Down and Inside Out," *Gnosis: Journal of Gnostic Studies* 1 (2016): 7–35.
Édouard Des Places, *Numenius. Fragments* (Paris: Les Belles Lettres, 1973).
Édouard Des Places, "Platon et la langue des mystères," in *Études platoniciennes, 1929–1979. Etudes préliminaires aux Religions orientales dans l'Empire romain*, 90 (Leiden: Brill, 1981), 83–98.
Franz J. Dölger, "'Sacramentum infanticidii.' Die Schlachtung eines Kindes und der Genuß seines Fleisches und Blutes als vermeintlicher Einweihungsakt im ältesten Christentum', in *Antike und Christentum* 4 (1934): 188–228.
Heinrich Dörrie, "Mysterien (in Kult und Religion) und Philosophie," in *Die Orientalischen Religionen im Römerreich*, ed. Maarten J. Vermasseren (Leiden: Brill, 1981), 341–62.
Heinrich Dörrie, "Philosophie und Mysterium," in *Verbum et Signum* II: *Beitrage zur mediavistischen Bedeutungsforschung Studien zu Semantik und Sinntradition im Mittelalter*, ed. Hans Fromm, Wolfgang Harms, and Uwe Ruberg (München: Fink, 1975), 9–24.
Garth Fowden, *The Egyptian Hermes. A Historical Approach to the late Pagan Mind* (Cambridge: Cambridge University Press, 1986).
Michael Frede, "Numenius," *Aufstieg und Niergang der römischen Welt* II.36.2 (1987): 1034–75.
Michael Frede, "Monotheism and Pagan Philosophy," in *Pagan Monotheism in Late Antiquity*, ed. Polymnia Athanassiadi and Michael Frede (Oxford; Oxford University Press, 1999), 41–68.
Konrad Gaiser, "Plato's Enigmatic lecture 'On the Good'," *Phronesis* 25 (1980): 5–37.
Dieter Georgi, "Weisheit Salomos," in idem, *Jüdische Schriften aus hellenistisch-römischer Zeit*, III.4 (Gütersloh: Mohn, 1980).
Marion Giebel, *Das Geheimnis der Mysterien. Antike Kulte in Griechenland, Rom und Ägypten* (Zürich – München: Patmos, 1990).
Christopher Gill, "Marcus Aurelius' *Meditations*: How Stoic and How Platonic?" in *Platonic Stoicism, Stoic Platonism*, ed. Mauro Bonazzi and Christoph Helmig (Leuven: Leuven University Press, 2007), 189–206.
Robert M. Grant, "Charges of 'Immorality' Against Various Religious Groups in Antiquity," in *Studies in Gnosticism and Hellenistic Religions Presented to G. Quispel on the Ocassion of his 65th Birthday*, ed. Roelof van den Broek and Maarten J. Vermaseren (Leiden: Brill, 1981), 161–70.
William K.C. Guthrie, *A History of Greek Philosophy*, 5: *The Later Plato and the Academy* (Cambridge: Cambridge University Press, 1978).
Wouter J. Hanegraaff et al., eds., *Dictionary of Gnosis and Western Esotericism* (Leiden: Brill, 2005).
Wouter J. Hanegraaff et al., eds., "Esotericism Theorized: Major Trends and Approaches to the Study of Esotericism," in *Religion: Secret Religion*, ed. April D. DeConick (San Francisco: MacMillan, 2016), 155–70.
Albert Henrichs, "Pagan Ritual and the Alleged Crimes of the Early Christians. A Reconsideration," in *Kyriakon. Festschrift J. Quasten I*, ed. Patrick Granfield, Josef Andreas Jungmann (Münster: Aschendorff, 1970), 18–35.
Albert Henrichs, *Die Phoinikika des Lollianos. Fragmente eines neuen griechischen Romans* (Bonn: R. Habelt, 1972).

Helmut Holzhey and Walther Ch Zimmerli, eds., *Esoterik und Exoterik der Philosophie: Beiträge zu Geschichte und Sinn philosophischer Selbstbestimmung: Rudolf W. Meyer zum 60. Geburtstag* (Basel; Stuttgart: Schwabe, 1977).

Brad Inwood, "Seneca, Plato and Platonism: the Case of Letter 65," *Platonic Stoicism and Stoic Platonism*, ed. Mauro Bonazzi and Christoph Helmig (Leuven: Leuven University Press, 2008), 149–67.

Irenaeus, *St. Irenaeus of Lyons: Against the Heresies*, ed. and translated by Dominic J. Unger, and John J. Dillon; Ancient Christian Writers 55, 64–65 (New York, N.Y.: Paulist Press, 1992).

Otto Kaiser, *Grundriss der Einleitung in die kanonischen und deuterokanonischen Schriften des Alten Testaments*, 3: *Die poetischen und weisheitlichen Werke* (Gütersloh: Gütersloher Verlagshaus, 1994).

Karen King, *What is Gnosticism* (Cambridge: Belknap Press, 2003).

Karen King, *The Secret Revelation of John* (Cambridge: Harvard University Press, 2006).

Peter Kingsley, *Ancient Philosophy, Mystery, and Magic. Empedocles and Pythagorean Tradition* (Oxford: Oxford University Press, 1995).

Hans G. Kippenberg and Guy G. Stroumsa, eds., *Secrecy and Concealment: Studies in the History of Mediterranean and Near Eastern Religions* (Leiden: Brill, 1995).

Hans Joachim Kraemer, *Arete bei Platon und Aristoteles. Zum Wesen und zur Geschichte der platonischen Ontologie* (Heidelberg: Carl Winter Universitätsverlag, 1959).

Hans Joachim Kraemer, *Plato and the Foundations of Metaphysics. A Work on the Theory of the Principles and Unwritten Doctrines of Plato with a Collection of the Fundamental Documents* (Trans. J. R. Catan; Albany: State University of New York Press, 1990).

Robert Lamberton, "The ἀπόρρητος θεωρία: the Roles of Secrecy in the History of Platonism," in *Secrecy and Concealment: Studies in the History of Mediterranean and Near Eastern Religions*, ed. Hans Kippenberg and Guy Stroumsa (Leiden: Brill, 1995), 139–52.

Pamela O. Long, *Openness, Secrecy, Authorship: Technical Arts and the Culture of Knowledge from Antiquity to the Renaissance* (Baltimore: Johns Hopkins University Press, 2001).

Tanya M. Luhrmann, "The Magic of Secrecy," *Ethos* 17 (1989): 131–65.

George W. MacRae, "Discourses of the Gnostic Revealer," in *Proceedings of the International Colloquium on Gnosticism. Stockholm, Aug. 20–25, 1973*, ed. Geo Windegren and David Hellholm (Stockholm and Leiden, 1977), 111–22.

Jean P. Mahé, "A Reading of the *Discourse on the Ogdoad and the Ennead* (Nag Hammadi Codex VI.6)," in *Gnosis and Hermeticism from Antiquity to Modern Times*, ed. Roelof van den Broek and Wouter J. Hanegraaff (Albany, N.Y.: State University of New York Press, 1988), 79–85. Jean P. Mahé, "La voie d'immortalité à la lumière des 'Hermetica' de Nag Hammadi et de découvertes plus," *Vigiliae Christianae* 45 (1991), 347–75.

Jaap Mansfeld, "Providence and the Destruction of the Universe in Early Stoic Thought. With Some Remarks on the 'Mysteries of Philosophy'," in *Studies in Hellenistic Religions*, ed. Maarten J. Vermaseren (Leiden: Brill, 1979), 129–88.

Giovanni Marasco, "Sacrifici umani e cospirazioni politiche," *Sileno* 7 (1981): 167–78.

Rainer Marten, "'Esoterik und Exoterik' oder 'Die philosophische Bestimmung wahrheitsfahiger Offentlichkeit', demonstriert an Platon und Aristoteles," in *Esoterik und Exoterik der Philosophie: Beiträge zu Geschichte und Sinn philosophischer Selbstbestimmung; Rudolf W. Meyer zum 60. Geburtstag*, ed. Helmut Holzhey (Basel: Schwabe, 1977), 13–31.

Luther H. Martin, "Secrecy in Hellenistic Religious Communities," in *Secrecy and Concealment: Studies in the History of Mediterranean and Near Eastern Religions*, ed. Hans Kippenberg and Guy Stroumsa (Leiden: E.J. Brill, 1995), 101–21.

Zeke Mazur, "Notes pour Plotin, Traite 33 (II 9) Contre les Gnostiques. Draft 1," in *Plotin: Oeuvres completes*. Tome 7, ed. Jean-Marc Narbonne, Mauricio Pagotto Marsola, Lorenzo Ferroni, Kevin Corrigan, and John D. Turner, Collection des Universités de France-Association Gillaume Bude (Paris: Les Belles Lettres, Forthcoming), 74–76.

Zeke Mazur, "Primordial Self-Reversion and the Gnostic Background of Plotinian Procession," Unpublished paper presented at the Annual Meeting of the International Society for Neoplatonic Studies. New Orleans, LA, June 2005.

Burkhard Meissner, "Mündliche Vermittlung und schriftliche Unterweisung in der antiken Berufsausbildung," in *Antike Fachschriftsteller: Literarischer Diskurs und sozialer Kontext*, ed. Marietta Horster and Christiane Reitz (Wiesbaden, 2003), 153–75.

Birgitta Nedelmann, "Geheimhaltung, Verheimlichung, Geheimniseinige Soziologische Vorüberlegungen," in *Secrecy and Concealment: Studies in the History of Mediterranean and Near Eastern Religions*, ed. Hans Kippenberg and Guy Stroumsa (Leiden: Brill, 1995), 1–16.

Martin Neher, *Wesen und Wirken der Weisheit in der Sapientia Salomonis* (Berlin: De Gruyter, 2004).

Arthur D. Nock and André J. Festurière, *Hermès Trismégiste. Corpus Hemerticum*, 4 vols (Paris: Les Belles Lettres, 1954–1960).

Jan Opsomer, "Plutarch on the One and the Dyad," in *Greek and Roman Philosophy 100 BC–200 AD*, ed. Robert W. Sharples and Richard Sorabji (London: Institute of Classical Studies, University of London, 2007), 379–95.

Jean Pepin, *Mythe et allégorie: Les origines grecques et les contestations judéo-chrétiennes* (Paris: Aubier, 1976).

Jean Pepin, "L'arcane religieux et sa transposition philosophique dans la tradition platonicienne," in *La storia della filosofia come sapere critico* (Milan: Franco Angeli editore, 1984), 18–35.

Apostolos L. Pierris, *Mystery and Philosophy: The Emergence of Reason from the Spirit of Mystery: An Inquiry into the Origin and Nature of Ancient Greek Rationality*, 4 vols (Patras: Institute for Philosophical Research, 2006–2007).

Giovani Reale, *Toward a New Interpretation of Plato* (Translated from the tenth edition and edited by John R. Catan and Richard Davies; Washington, DC: The Catholic University of America Press, 1997).

James M. Reese, *Hellenistic Influence on the Book of Wisdom and Its Consequences* (Rome: Biblical Institute Press, 1970).

Christoph Riedweg, *Mysterienterminologie bei Platon, Philon und Klemens von Alexandrien* (Berlin: DeGruyter, 1987).

John M. Rist, "Are You A Stoic? The Case of Marcus Aurelius," in *Jewish and Christian Self-Definition, 3: Self-Definition in the Graeco-Roman World*, ed. Ben F. Meyer and Ed P. Sanders (London: SCM, 1982), 23–45.

Alexander Roberts and James Donaldson, *The Ante-Nicene Fathers*, revised and chronologically arranged, with brief prefaces and occasional notes by A. Cleveland Coxe (Edinburgh: T&T Clark, 1885).

Lautaro Roig Lanzillotta, "The Early Christians and Human Sacrifice," *The Strange World of Human Sacrifice*, ed. Jan N. Bremmer (Leuven: Peeters, 2007), 81–102.

Lautaro Roig Lanzillotta, "Dios como Padre y artífice en *Moralia* de Plutarco," in *Filiación. Cultura pagana, religión de Israel, orígenes del cristianismo*, V, ed. Patricio de Navascués, Manuel Crespo, Andrés Sáez (Madrid: Trotta, 2013), 139–56.

Lautaro Roig Lanzillotta, "The Cosmology of the *Ascension of Isaiah*: Analysis and Re-Assessment of the Text's Cosmological Framework," *The Ascension of Isaiah*, ed. Jan N. Bremmer, Thomas R. Karmann, Tobias Nicklas (Leuven: Peeters, 2015), 259–88.

Lautaro Roig Lanzillotta, "La recepción de Platón, *Timaeus* 28C, in Clemente de Alejandría," *Filiación. Cultura pagana, religión de Israel, orígenes del cristianismo*, VI, ed. Patricio de Navascués, Manuel Crespo, Andrés Sáez (Madrid: Trotta, 2016), 259–80.

Lautaro Roig Lanzillotta, "The Apocalypse of Paul (NHC V,2): Cosmology, Anthropology, and Ethics," *Gnosis: Journal of Gnostic Studies* 1 (2016).

Lautaro Roig Lanzillotta, "Plutarch and the Image of the Sleeping and Waking Soul," in *Immagini letterarie e iconografia nelle opere di Plutarco, Madrid, Università di Salerno/ Red Temática Europea "Plutarco,"* ed. Stefano Amendola, Giovanna Pace and Paola Volpe Cacciatore (Madrid: Ediciones Clásicas, 2017), 209–22.

Lautaro Roig Lanzillotta, "Spirit, Soul and Body in Nag Hammadi Literature: Distinguishing Anthropological Schemes in Valentinian, Sethian, Hermetic and Thomasine Texts," *Gnosis: Journal of Gnostic Studies* 2 (2017), 15–39.

Lautaro Roig Lanzillotta, "Achamot, el Alma del mundo valentiniana, y su relación con el Demiurgo (Ireneo, Adv. Haer. 1.5)," in *Jornadas sobre la Filiación*, VII, ed. Patricio Navascués and Andrés Sáez (Madrid: Trotta, 2018). Forthcoming.

Lautaro Roig Lanzillotta, "Knowledge's Silent Stream: Esotericism from the Presocratics to the Gnostics," 4–11. (Forthcoming).

Kurt Rudolph "Geheimnis und Geheimhaltung in der antiken Gnosis und im Manicheïsmus," in *Secrecy and Concealment: Studies in the History of Mediterranean and Near Eastern Religions*, ed. Hans Kippenberg and Guy Stroumsa (Leiden: Brill, 1995), 265–87.

Barbara M. Sattler, "The Eleusinian Mysteries in Pre-Platonic Thought," in *Philosophy and Salvation in Greek Philosophy*, ed. Vishwa Adluri (Göttingen: De Gruyter, 2013), 151–90.

Renate Schlesier, "Maskierter Texte. Religiose Anspielung und Verheimlichung in der griechischen Tragödie," in *Secrecy and Concealment: Studies in the History of Mediterranean and Near Eastern Religions*, ed. Hans Kippenberg and Guy G. Stroumsa (Leiden: Brill, 1995), 123–38.

Eduard Schweizer, "Die hellenistische Komponente im neutestamentlichen sarx-Begriff," *Zeitschrift für die neutestamentliche Wissenschaft* 48 (1957): 237–53.

Walter Scott, *Hermetica: the Ancient Greek and Latin Writings which Contain Religious or Philosophic Teachings Ascribed to Hermes Trismegistus* (Oxford: Oxford University Press 1926–1934).

David Sedley, "The Origins of Stoic God," in ed. Dorothea Frede and André Laks (Leiden / Boston: Brill), 41–84.

Georg Simmel, "The Sociology of Secrecy and of Secret Societies," *American Journal of Sociology* 11 (1906): 441–98.

Georg Simmel, *The Sociology of Georg Simmel* (transl. K. Wolff; Glencoe, Ill: Free Press, 1950).

Jonathan Z. Smith, *Differential Equations: On Constructing the 'Other'* (Thirteenth Annual University Lectures in Religion; Arizona State University, Department of Religious Studies, 1992).

Guy G. Stroumsa, "From Esotericism to Mysticism in Early Christianity," in *Secrecy and Concealment: Studies in the History of Mediterranean and Near Eastern Religions*, ed. Hans Kippenberg and Guy Stroumsa (Leiden: Brill, 1995), 289–309.

Guy G. Stroumsa, *Hidden Wisdom: Esoteric Traditions and the Roots of Christian Mysticism* (Leiden: Brill, 1996).

Szlezák, T.A., "Plato," (section E, "Criticism of writing"), in *Brill's New Pauly: Antiquity volumes*, ed. Hubert Cancik and Helmuth Schneider. Brill Online, 2014. Reference. 08 February 2014.

Stanton K. Tefft, *Secrecy: A Cross-Cultural Perspective* (New York: Human Sciences Press, 1980).

Eugène N. Tigerstedt, *The Decline and Fall of the Neoplatonic Interpretation of Homer* (Helsinki: Societas Scientiarum Fennica, 1974).

Eugène N. Tigerstedt, *Interpreting Plato* (Stockholm: Almqvist and Wiksell International, 1977).

Laurence M. V. Totelin, *Hippocratic Recipes: Oral and Written Transmission of Pharmacological Knowledge in Fifth- and Fourth-Century Greece* (Leiden: Brill, 2009).

John D. Turner, "Transgressing Boundaries: Plotinus and the Gnostics," *Gnosis: Journal of Gnostic Studies* 1 (2016): 56–85.

Hugh B. Urban, "The Torment of Secrecy: Ethical and Epistemological Problems in the Study of Esoteric Traditions," *History of Religions* 37 (1998): 209–48.

Heinrich von Staden, "Body, Soul, and Nerves: Epicurus, Herophilus, Erasistratus, the Stoics, and Galen," in *Psyche and Soma: Physicians and Metaphysicians on the Mind-Body Problem From Antiquity to Enlightenment*, John Wright and Paul Potter (Oxford: Oxford University Press, 2003), 79–116.

Martin West, "Towards Monotheism," in *Pagan Monotheism in Late Antiquity*, Polymnia Athanassiadi and Michael Frede (Oxford: Oxford University Press, 1999), 21–40.

Michael A. Williams, "Secrecy, Revelation, and late Antique Demiurgical Myths," in *Rending the Veil: Concealment and Secrecy in the History of Religions*, ed. Elliot R. Wolfson (New York / London: Seven Bridges Press, 1999), 31–58.

Elliot R. Wolfson, ed., *Rending the Veil: Concealment and Secrecy in the History of Religions* (New York/ London: Seven Bridges Press, 1999).

George van Kooten
7 The Sign of Socrates, the Sign of Apollo, and the Signs of Christ: Hiding and Sharing Religious Knowledge in the Gospel of John – A Contrapuntal Reading of John's Gospel and Plato's Dialogues

7.1 Introduction

As I am currently arguing elsewhere, it seems very likely that John's Gospel, in its combination of the genres of ancient biography and dialogue, shows itself acquainted with the Platonic dialogues, and particularly with the *Symposium* and the dialogues describing the last days of Socrates.[1] Not only does Plato's *Symposium* serve to reinterpret Jesus's last supper, but its sympotic model, often combined with its subject matter of love and generation, is applied throughout the gospel. Borrowing Kathryn Topper's differentiation between "symposia of the primitive," with people reclining directly on the ground without furniture, and luxurious symposia (a differentiation derived from Plato's *Republic* 2.13, 372B–373A),[2] it is possible to recognise these two types throughout the Gospel's narrative. The "symposia of the primitive" include such sympotic events as the meeting with the Samaritan woman – insatiable for amorous love – at the well, where water and bread remain untouched (John 4); the feeding of the 5,000, where people recline directly on the ground (John 6); and the primitive post-resurrection symposium on the shores of the Sea of Tiberias, where denied love is restored (John 21). Examples of the luxurious symposia are the wedding of Cana with its nuptial love, and an ἀρχιτρίκλινος in charge of the festivities (John 2); the luxurious symposium at the house of Lazarus with richly provided myrrh

---

1 For the latter, see George van Kooten, "The Last Days of Socrates and Christ: *Euthyphro*, *Apology*, *Crito*, and *Phaedo* Read in Counterpoint with John's Gospel," in *Religio-Philosophical Discourses in the Mediterranean World: From Plato, through Jesus, to Late Antiquity*, ed. Anders K. Petersen and George van Kooten, Ancient Philosophy & Religion 1 (Leiden: Brill, 2017), 219–43.
2 Kathryn Topper, *The Imagery of the Athenian Symposium* (Cambridge: Cambridge University Press, 2012), 23–52.

Note: I dedicate this piece to the memory of Jan Schaap (1936–2015), my old headmaster at the Prins Maurits School in Delft, whose funeral coincided with the conference. I remember him with great gratitude, also for his strong interest and support over the years.

(John 12); and the Last Symposium with its new commandment of mutual love (John 13–17); not to mention the heavenly symposium with the only-begotten Son reclining intimately with the divine Father (John 1:18), in extreme contrast with Jesus finally "reclining" his head on the cross after he has drunk the wine-vinegar (John 19:30, … ἔλαβεν τὸ ὄξος … καὶ κλίνας τὴν κεφαλήν …).

In addition to Plato's *Symposium*, also his dialogues describing the last days of Socrates seem to have been very relevant for John, especially the *Euthyphro*, the *Apology*, and the *Phaedo*. The former two, with their attention for the charges against Socrates – the charges of making new gods, including his own divine *daimonion* – seem to resonate with John's Gospel in its depiction of the charges against Jesus, of making himself god. The *Phaedo*, with its full focus on immortality and its reference to the eschatological vision of "the true light" after one's death (109E), is echoed in John's attention for eternal life and the true light (John 1:9; 1 John 2:8), whilst its statements about what is truly real, as opposed to what "one can touch and see and drink and eat" (65D–66B; 81B; 84A–B),[3] are reflected throughout John's Gospel, particularly in sympotic settings where true food and true drink are contrasted with merely physical sustenance.[4]

In my contribution to the present volume, I want to focus on a further symmetry between Plato and John, in their interest for, respectively, the "sign" of Socrates and the "signs" of Christ. Signs, because of their semiotic ambiguity of signifying references to another, signified reality are encapsulated in the dynamics of sharing and hiding, the theme of this volume on "sharing" and "hiding" religious knowledge. Insofar as signs refer to another reality they "share" information; at the same time, however, because of the subtlety of signs they conceal and "hide" at the same time. In search of a better understanding of John's semiotic play of signification, I will take my starting point in Plato's dialogues describing the last days of Socrates, in which he pays much attention to the so-called "Sign of Socrates." First, I will discuss Plato's use of this notion in these and other dialogues and emphasise its apotreptic nature – its

---

[3] As regards translations, classical authors are normally quoted from the Loeb Classical Library and the New Testament writings from the New Revised Standard Version, with adaptations where necessary or useful.
[4] For a comparison between John's Gospel and Plato's dialogues about the last days of Socrates, see van Kooten, "The Last Days of Socrates and Christ." For a comparison of John's "true light" with that of Plato, see George van Kooten, "The 'True Light which Enlightens Everyone'" (John 1:9): John, Genesis, the Platonic Notion of the "True, Noetic Light," and the Allegory of the Cave in Plato's *Republic*," in *The Creation of Heaven and Earth: Re-interpretations of Genesis I in the Context of Judaism, Ancient Philosophy, Christianity, and Modern Physics*, ed. George van Kooten, Themes in Biblical Narrative 8 (Leiden: Brill, 2005), 149–94.

characteristic of withholding Socrates from particular actions for a particular time until the right moment has come to resume or perform his intended activities. Secondly, I will comment on this sign as grounded in Socrates's Apollonian mission, so that Socrates's sign can be seen as a manifestation of Apollo's oracular prophetic power. Subsequently I will compare these features of Socrates's sign with John's portrayal of Christ's signs, divine mission, and often equally apotreptic behaviour. Fourthly and finally, I will compare Apollo's "signification', as understood by Platonists such as Plutarch, with John's semiotic programme as outlined in his Gospel. Plutarch, in his *De Pythiae oraculis*, reflects extensively on Heraclitus's characterization of Apollo as "The Lord whose prophetic shrine is at Delphi neither tells nor conceals, but signifies" (B 93 DK). This Heraclitean axiom explicitly expresses the view that Apollo's signs neither tell nor conceal; they simultaneously share and hide religious knowledge, and this is exactly what signifying is. I will end by concluding that this understanding of religious signs can illuminate our understanding of the semiotic programme of John's Gospel. It seems that John, in his Gospel, places himself in continuity with this Greek semiotic discourse.

7.2 The Sign of Socrates

Socrates's "sign" (σημεῖον) is mentioned in Plato's dialogues describing the last days of Socrates, but continues to feature throughout the other dialogues. It is often called Socrates's "daimonion" (δαιμόνιον), and sometimes the "voice" (φωνή) which occurs to him. In the *Euthyphro*, Socrates's reference to his *daimonion* is taken as the reason why Socrates was charged with introducing new *daimonia*. When Socrates tells Euthyphro that Meletus of Pitthus has indicted him for making and introducing new gods, at the expense of the old gods, Euthyphro immediately links this charge with Socrates's frequent mentioning of his own *daimonion*:

> SOCRATES: ... he says I am a maker of gods; and because I make new gods and do not believe in the old ones, he indicted me for the sake of these old ones, as he says.
>
> EUTHYPHRO: I understand, Socrates; it is because you say the *daimonion* keeps coming to you (ὅτι δὴ σὺ τὸ δαιμόνιον φῂς σαυτῷ ἑκάστοτε γίγνεσθαι). So he has brought the indictment against you for making innovations in religion ... (Plato, *Euthyphro* 3B)

In his defence against these charges in the *Apology*, Socrates is said to confirm this link, as he explicitly says that Meletus, in his indictment, is targeting his frequent reference to his *daimonion*, the possession of which, Socrates tells the court, prohibits him from engaging in politics but lets him interact with people's private lives:

> Perhaps it may seem strange that I go about and interfere in other people's affairs to give this advice in private, but do not venture to come before your assembly and advise the state. But the reason for this, as you have heard me say at many times and places, is that something divine and spiritual comes to me, the very thing which Meletus ridiculed in his indictment (τούτου δὲ αἴτιόν ἐστιν ὃ ὑμεῖς ἐμοῦ πολλάκις ἀκηκόατε πολλαχοῦ λέγοντος, ὅτι μοι θεῖόν τι καὶ δαιμόνιον γίγνεται [φωνή], ὃ δὴ καὶ ἐν τῇ γραφῇ ἐπικωμῳδῶν Μέλητος ἐγράψατο). I have had this from my childhood (ἐμοὶ δὲ τοῦτ' ἔστιν ἐκ παιδὸς ἀρξάμενον); it is a sort of voice that comes to me, and when it comes it always holds me back from what I am thinking of doing, but never urges me forward (ἀεὶ ἀποτρέπει με τοῦτο ὃ ἂν μέλλω πράττειν, προτρέπει δὲ οὔποτε). This it is which opposes my engaging in politics. (Plato, *Apology* 31D)

Hence, according to Plato there is indeed a direct connection between Socrates's claim to possess a specific *daimonion* and the charges against him for introducing "new *daimonia* (δαιμόνια καινά)" (*Apology* 24B–C, 26B; cf. *Euthyphro* 3B), a view shared by Xenophon (see his *Memorabilia* 1.1.1–4). The charge of having introduced new *daimonia* is levelled against Socrates's frequent references to his own *daimonion*. According to Plato, Socrates had repeatedly drawn the Athenians' attention to this *daimonion*, "at many times and places." And the nature of this *daimonion* is "apotreptic," not "protreptic": it holds him back (ἀποτρέπει) from what he is thinking of doing, but never urges him forward (προτρέπει δὲ οὔποτε). Later in his apologetic defence before the court, Socrates is said to return to this issue of the normally apotreptic nature of his *daimonion*, now also designating it as his "sign (σημεῖον)," and arguing that its cessation implies that the hour of his departure through death has drawn near. As Socrates tells the court,

> ... a wonderful thing has happened to me. For hitherto the customary prophetic skill of the *daimonion* (ἡ γὰρ εἰωθυῖά μοι μαντικὴ ἡ τοῦ δαιμονίου) always spoke to me very frequently and opposed me even in very small matters, if I was going to do anything I should not; but now, as you yourselves see, this thing which might be thought, and is generally considered, the greatest of evils has come upon me; but the sign of the god (τὸ τοῦ θεοῦ σημεῖον) did not oppose me either when I left my home in the morning, or when I came here to the court, or at any point of my speech, when I was going to say anything; and yet on other occasions it stopped me at many points in the midst of a speech; but now, in this affair, it has not opposed me in anything I was doing or saying. What then do I suppose is the reason? I will tell you. This which has happened to me is doubtless a good thing, and those of us who think death is an evil must be mistaken. A convincing proof of this has been given me; for the accustomed sign (τὸ εἰωθὸς σημεῖον) would surely have opposed me if I had not been going to meet with something good. (Plato, *Apology* 40A–C)

Socrates's *daimonion* is now also identified as "the sign of the god" (τὸ τοῦ θεοῦ σημεῖον) or, to stress its repetitive manifestation, "the accustomed sign (τὸ εἰωθὸς σημεῖον)"; and its usually apotreptic nature, stopping Socrates "at many points in the midst of a speech," is confirmed. At the very end of his speech, Socrates emphasises again that the fact that the sign has not been apotreptic during the trial indicates that "the hour has come to go away," and that, since he is dying as a good man, his death is not accidental but at the right, appropriate time, showing God's care for him:

> ... no evil can come to a good man either in life or after death, and God does not neglect him. So, too, this which has come to me has not come by chance, but I see plainly that it was better for me to die now and be freed from troubles. That is the reason why the sign never deterred me (διὰ τοῦτο καὶ ἐμὲ οὐδαμοῦ ἀπέτρεψεν τὸ σημεῖον), and I am not at all angry with those who condemned me or with my accusers. ... But now the time ("hour") has come to go away (ἀλλὰ γὰρ ἤδη ὥρα ἀπιέναι). I go to die, and you to live; but which of us goes to the better lot, is known to none but God. (Plato, *Apology* 41C–D, 42A)

Let me already draw attention to the co-occurrence here of the idea of (the ceasing) of the apotreptic behaviour of Socrates's divine sign and the notion of the appropriate "hour," which seems to indicate the underlying concept of a divine chronology which permeates Socrates's life from his childhood (cf. *Apology* 31D) right up until his death. We will encounter a similar divine chronological regulation in our discussion of John's Gospel, where we find the same interrelatedness between apotreptic delay and resumption of action at the proper, divine hour and time.

The issue of the apotreptic conduct of Socrates's daemonic sign is not confined to dialogues describing the last days of Socrates but also occurs in the *Symposium* and is also important in other Platonic dialogues. It is useful to notice the concrete circumstances in which this sign apotrepically delays Socrates's actions. When, according to the *Symposium*, Socrates is on his way to a dinner party at the house of Agathon, and takes Aristodemus along as an uninvited guest (*Symposium* 174A–B), it is Socrates who falls behind because of his apotreptic *daimonion*, and sends his companion ahead to arrive first at Agathon's house, without Socrates (174D–E). The reason for Socrates's delay is that he becomes "absorbed in his own thoughts by the way" (174D), even to such an extent that the slave who is sent out by Agathon to fetch him returns with the news that "this Socrates had retreated into their neighbours' porch; there he was standing, and when bidden to come in, he refused" (175A).[5] When Agathon orders his servant to continue bidding Socrates to come in, Aristodemus interferes in a way very similar – if I may add another brief comparison with John's Gospel – to the way Jesus's mother Mary responds to the servants of the ἀρχιτρίκλινος when Jesus still apotreptically

---

[5] Cf. also the story of Socrates's standing outside in the cold from dawn till dawn when on a campaign in *Symposium* 219E–220D.

refrains from wine-making at the wedding at Cana (John 2:3–5): just as Mary tells the servants to be patient but alert, so Aristodemus tells Agathon not to press Socrates but to remain calm as he will show up at his own, appropriate time: "let him alone; it is a habit he has. Occasionally he turns aside, anywhere at random, and there he stands. He will be here presently, I expect. So do not disturb him; let him be" (175B). Socrates does indeed finally arrive, but not before the dinner part of the symposium is well underway; and before his arrival Aristodemus has frequently had to restrain Agathon's impulses to force Socrates inside: "they all began dinner, but Socrates did not arrive; and though Agathon ever and anon gave orders that they should go and fetch him, [Aristodemus] would not allow it. When he did come, it was after what, for him, was no great delay, as they were only about half-way through dinner" (175C). Again, the remark that Socrates is so seriously delayed that he only arrives "half-way" seems to parallel the event in John's Gospel where Jesus does not allow his brothers to pressurise him to come to a particular Jewish festival in Jerusalem, but does so in his own time, only displaying himself "about the middle of the festival" (John 7:2–14).

As is apparent from these examples, both Plato and John depict this apotreptic behaviour of Socrates and Christ in very normal, daily circumstances. It seems that both Socrates and Christ respond to a higher calling, and take deliberate time for reflection, and wait till their divine chronology is synchronised with the mundane, daily timetables of their fellow human beings. In my understanding, the most convincing way to explain John's apotreptic portrayal of Jesus in his Gospel is to assume that John was indeed acquainted with Plato's dialogues.

Apart from the *Apology* and the *Symposium*, Socrates's apotreptic sign is also mentioned in other Platonic dialogues as characteristic of Socrates's behaviour. In the *Phaedrus* Socrates tells Phaedrus that when he was about to cross a stream "the spirit and the sign that usually comes to me (τὸ δαιμόνιόν τε καὶ τὸ εἰωθὸς σημεῖόν μοι γίγνεσθαι)," which "always holds me back from something I am about to do (ἀεὶ δέ με ἐπίσχει ὃ ἂν μέλλω πράττειν)," prohibited him from crossing; he was not permitted to continue until he had cleared his conscience and had come to realise that he had "committed some sin against deity" through a speech which he had just delivered and which was "foolish, and somewhat impious" (*Phaedrus* 24B–D).

According to the *Euthydemus*, in different, but equally everyday circumstances "the customary daemonic sign (τὸ εἰωθὸς σημεῖον τὸ δαιμόνιον)" prevents Socrates from getting up in the undressing-room of the Lyceum's gymnasium and leaving (*Euthydemus* 272E–373A). This enables him to enter into critical debate with two sophists, Euthydemus (who has lent the dialogue its name) and Dionysodorus, on the important issue of virtue and true education, about which Socrates and the sophists entertain such different views. Indeed, as Socrates explains in book VII of the *Republic*, it is "the daemonic sign (τὸ δαιμόνιον σημεῖον)" that ensures that he belongs to the "very small remnant (...) of those who consort worthily with philosophy" (496A–C) and are not affected by the degeneracy of contemporary politics (496C). Some people, according to Socrates, remain faithful to philosophy by being "held in check by exile," which prevents them from being corrupted; others just detest the parochialism of their environment, or are unsatisfied by the other arts and feel a natural affinity towards philosophy, or are luckily kept out of demoralising politics by a sickly constitution. His own dedication to philosophy, however, is due to "the daemonic sign (τὸ δαιμόνιον σημεῖον)," something which "is hardly worth mentioning," he says, "for I suppose it has happened to few or none before me" (496C). Socrates's devotion to philosophy is thus regarded as the result of a unique, divine calling and mission, which – as we will see in Section 3 of this paper – can perhaps best be understood as an Apollonian calling.

Finally, in the *Theaetetus*, Socrates's *daimonion* informs him about the very concrete issue of which of his former pupils, who have prematurely left him but afterwards regret having done so, are still "supervisable":

> When such men come back and beg me, as they do, with wonderful eagerness to let them join me again, the *daimonion* that comes to me (τὸ γιγνόμενόν μοι δαιμόνιον) forbids me to associate with some of them, but allows me to converse with others, and these again make progress (Plato, *Theaetetus* 151A).[6]

It thus appears that Socrates's divine sign displays itself in very concrete circumstances, but is nevertheless always tuned to the bigger issues of virtue, true education, and pious speech; lurking in the background is the strife between philosophers on the one hand, and sophists and demoralised politicians on the other. Furthermore, it also seems that although Socrates's sign is rightly depicted as apotreptic, as it dissuades him from particular actions, this apotreptic delay is mainly only temporary, allowing Socrates time for much-needed reflection before resuming action, or granting him the opportunity to engage in a particular debate. Hence, the apotreptic nature of Socrates's sign is not simply

---

6 See further, among Plato's spurious writings, [*Alcibiades major*] 103A–B and *Theages* 129B–D; also here, Socrates's divine sign – called "a certain daemonic opposition (τι δαιμόνιον ἐναντίωμα)" ([*Alc. maj.*] 103A) or "the accustomed daemonic sign (τὸ εἰωθὸς σημεῖον τὸ δαιμόνιον)" (*Theages* 129B), or simply "the sign (τὸ σημεῖον)" (129D) – is apotreptic and reveals itself in very concrete circumstances. Such circumstances include dissuading Socrates from meeting Alcibiades because he is such a self-confident person ([*Alc. maj.*] 103A–104A), trying to prevent someone from committing a murder (*Theages* 129A–D), and other specific circumstances (*Theages* 129D) and discussions (*Theages* 129E).

negative; rather it serves to develop a particular reflective and receptive attitude that keeps pace with a divine timing, which breaks the daily grind of earthly patterns or inclinations. This sign seems to make Socrates reflective, and also receptive and attentive to opportunities which present themselves, so that, after an apotreptic pause, he is often able to resume or redirect his actions in a different, or deeper sense and awareness. Given the complex operation of this divine sign, it is understandable that Platonists such as Plutarch noticed that it is not simply apotreptic, but combines apotreptic and protreptic features (*De genio Socratis* 581A–B).

7.3 The Sign of Apollo

Socrates's sign is not only apotreptic, it is also divine, and therefore directly relevant for this volume's focus on sharing and hiding religious knowledge. It is a "daemonic," divine sign, "the sign of the god (τὸ τοῦ θεοῦ σημεῖον)," as Plato calls it in *Apology* 40B. This of course prompts the question which god Plato had particularly in mind. But prior to that we should perhaps raise the preliminary question of whether Plato was even necessarily thinking of a god at all. Plutarch, for instance, mentions two very different interpretations of Socrates's sign: a reductionist interpretation which suggests that Socrates was actually referring to his intellect, which he had in common with all other human beings; and another interpretation which assumes that Socrates was did indeed referring to a divinity. As Plutarch puts it in his *Platonic Questions*:

> Is it then his own nature (φύσις) (...), that he [i.e., Socrates] called 'god' (θεός), as Menander said "for our intelligence is god" (ὁ νοῦς γὰρ ἡμῶν ὁ θεός) and Heraclitus "the character of a man is his guardian spirit" (ἦθος ἀνθρώπῳ δαίμων); or did some truly divine and spiritual cause guide Socrates to this kind of philosophy (ἢ θεῖόν τι καὶ δαιμόνιον ὡς ἀληθῶς αἴτιον ὑφηγήσατο Σωκράτει τοῦτο τῆς φιλοσοφίας τὸ γένος)...?" (999D–E)

The former view could perhaps be supported by Plato's view in his *Cratylus* that all good men have a *daimonion* (*Cratylus* 398B–C). Somehow, however, Socrates's *daimonion* seems to be rather unique, as we have seen him stating in the *Republic* (see the passage 496C quoted above). It seems more likely that the latter view is true, and that "the sign of the god (τὸ τοῦ θεοῦ σημεῖον)" refers to a real god, and not merely to the human intelligence possessed by good men. Although Plato does also call Socrates's divine sign a *daimonion*, this does not mean that Plato ascribed the origins of this sign to a *daimōn* rather than to a god. As Anthony A. Long has argued,

Did Socrates take his divine sign to be the voice of a *daimōn*? No firm answer to this question can be given, but the word *daimonion* (literally 'divine thing') is just as appropriate to a visitation from a fully fledged god, in which case its most likely source for Socrates would be the god Apollo whose oracle initiated his interpretation of his mission to the citizens of Athens.[7]

It does indeed seem very likely that "the sign of the god (τὸ τοῦ θεοῦ σημεῖον)" (*Apology* 40B) which Socrates possesses is the sign of Apollo. Among Plato's dialogues it is only in the *Apology* and the *Phaedo* that Plato reflects on Socrates's relation to Apollo. In *Apology* 20D–23C Plato describes Socrates's divine mission to the Athenians on behalf of that god. According to Socrates, his wisdom was confirmed by Apollo himself when Chaerephon, who was Socrates's comrade from his youth (20E–21A) and had apparently noticed his wisdom early on, asked Apollo's oracle at Delphi whether there were anyone wiser than Socrates, to which "the Pythia replied that there was no one wiser" (*Apology* 21A). Having had his wisdom divinely authorised, Socrates subsequently tries to prove the oracle wrong and to explore in which sense he could possibly be the wisest of men. By engaging with three classes – the politicians (οἱ πολιτικοί), the poets, and the hand-workers – he finds that many of them believe themselves wise and think they know something, or often know none of the things they say, or overestimate their own importance. In this way Socrates comes to realise that it is Apollo who "is really wise and by his oracle means this: 'Human wisdom is of little or no value'" (23A), and that those human beings are the wisest who, like Socrates himself, recognise that they are "in truth of no account in respect to wisdom" (23A–B). Hence, Socrates understands it as his divine mission to question the self-confessed wisdom of human beings "at the god's behest (κατὰ τὸν θεὸν)," "giving aid to the god (τῷ θεῷ βοηθῶν)" by showing that such human beings are not wise (23B). There is an unmistakably anti-sophistic ring to Socrates's mission as he states that he is "in vast poverty ... on account of this service to the god (ἐν πενίᾳ μυρίᾳ ... διὰ τὴν τοῦ θεοῦ λατρείαν)" (23B–C), having received no pay for his exhortations but, as his poverty shows beyond doubt (31B–C), offering himself "alike to rich and poor" (33B).[8] This is highly relevant to the theme of this volume,

---

[7] See Anthony A. Long, "daimōn," in *The Continuum Companion to Plato*, ed. Gerald A. Press (London: Continuum, 2012), 152–54.
[8] Socrates's criticism of the sophists and his own different mode of operation is rendered explicit by Plutarch, *Platonic Questions* I 999E–F, who depicts Socrates as "trustworthy" (ἀξιόπιστος) in contrast with the sophists. Cf. Socrates's interrogation of Callias, "a man who has spent more on sophists than all the rest" (*Apology* 20A), which precedes the discussion of Socrates's own wisdom in 20D–23C.

as it shows that it was Socrates's intention, and mission, to *reach out* to all social classes, "alike to rich and poor," and *to share* his insights.

Hence, Socrates finds the truth of Apollo's oracular statement that he is the wisest of all men confirmed and, at his trial for alleged impiety, is able to appeal to Apollo as a witness: "For of my wisdom – if it is wisdom at all – and of its nature, I will offer you the god of Delphi as a witness (τῆς γὰρ ἐμῆς, εἰ δή τίς ἐστιν σοφία καὶ οἵα, μάρτυρα ὑμῖν παρέξομαι τὸν θεὸν τὸν ἐν Δελφοῖς)" (*Apology* 20E). Plato thus closely connects Socrates's entire mission to the Athenians with Apollo, and it is clearly implied that Socrates's wisdom is the result of his possession of the sign of the God, which he had had "from childhood" (ἐκ παιδός, 31D), as was already recognised by his comrade Chaerephon (20E–21A). "The sign of the god" therefore must be Apollo's.

After establishing such a close connection between Socrates and Apollo in his description of Socrates's mission in the *Apology*, Plato, in his *Phaedo*, closely links Socrates and Apollo again by contextualising Socrates's death in the setting of the annual Athenian festival of Apollo. The beginning of the festival had taken place the day before Socrates's trial and was marked by particular rituals performed by the priest of Apollo, and although the death sentence had already been passed, Socrates's execution had to wait until the end of the festival several days later because during the festival no one could be publicly executed (*Phaedo* 58B–C). Interestingly, we find the same constraints of a religious festival upon the public execution of a convict emphasised in John's Gospel (and this element is absent from the other, Synoptic gospels). The Jews are said to arrive at Pilate's headquarters early in the morning but not to enter, "so as to avoid ritual defilement and to be able to eat the Passover" (John 18:28), and the final phase of the crucifixion is rushed so that it can be completed before the onset of the festival (19:31–34). In the case of Socrates his execution is delayed until after the festival. It is during these days between sentence and execution that Socrates, in breach of his life-long habit of not writing anything, is said to have composed, among other works, a hymn to Apollo (60D), addressed "to the god whose festival it was" (61B). – In passing I note that this extraordinary event of Socrates taking to writing, although none of these written compositions survive, seems paralleled in John's narrative when Jesus, in an equally or perhaps even more ephemeral mode, is said to have written in the sand (John 8:6 and 8:8, under a similar threat of indictment),[9] as inaccessible for posterity as Socrates's hymn to Apollo. Moreover, on account of this close bond between Socrates and Apollo,

---

[9] The authenticity of the passage of John 7:53–8:11 is text-critically uncertain, but its remark on Jesus's literacy would fit well with John's portrayal of Jesus's learning (7:15).

according to Plato's *Phaedo* Socrates also compares himself to the swans, which are considered "Apollo's birds," and states that they, "when they feel that they are to die, sing most and best in their joy that they are to go to the god whose servants they are" (84E–85B). Similarly, Socrates, too, foresees his own death[10]: "And I think that I am myself a fellow-servant of the swans, and am consecrated to the same God (καὶ ἱερὸς τοῦ αὐτοῦ θεοῦ) and have received from our master (δεσπότης) a gift of prophecy no whit inferior to theirs, and that I go out from life with as little sorrow as they" (*Phaedo* 85B). Hence, not only has Socrates received the sign of Apollo from childhood on, and has his wisdom been confirmed by Apollo's Delphic oracle, but he himself is also consecrated to Apollo, writes a hymn to Apollo in his final days, and departs at the end of Apollo's festival.

It seems that the specific choice of Apollo as Socrates's god is grounded in the following. First of all, Apollo is regarded as the inspirator and inventor of prophetic, oracular divination (ἡ μαντική) (*Phaedrus* 265B; *Symposium* 197A; *Cratylus* 405A and 405C), which seems to directly link up with the prophetic power of Socrates's *daimonion* (*Apology* 40A, ἡ ... μαντικὴ ἡ τοῦ δαιμονίου); at the same time, "education owes its origin to Apollo and the Muses" (*Laws* 654A). Moreover, the well-known philosophical inscriptions at Delphi of "Know thyself" and "Nothing overmuch," although assembled by the Spartans and dedicated by them "to Apollo in his Delphic temple" (*Protagoras* 343A–B), are also ascribed to the Pythian oracle (*Laws* 923A) and hence, in a sense, to Apollo himself.[11] Most importantly, perhaps, Plato regards Apollo as "the God of our fathers," who is "for all mankind the interpreter of the religion of their father" (*Republic* 427B–C). According to Plato all religious legislation of Athens resides with Apollo.[12]

---

10 In Plato's reasoning Socrates's foreknowledge of his death is probably the result of the fact that the apotreptic sign of Apollo no longer opposes him during the trial (*Apology* 40A–C; 41C–D). For further reflections on the relation between Socrates's *daimonion* and his ability to foresee the future, see Plutarch, *De genio Socratis* 581D-E.
11 For Plato's reference to the Delphic inscriptions, see also *Philebus* 48C; and cf. also [*Alcibiades major*] 124A, *Hipparchus* 228D-E, [*Amatores*] 138A and *Charmides* 164D-E. About the potential link between Socrates's spirituality and that of Delphi, see Mark L. McPherrran, *The Religion of Socrates* (University Park, Pennsylvania: Pennsylvania State University Press, 1996), 216–18.
12 For Apollo's Delphic oracles as the authorising institute of Athens's religious legislation, see also Plato's *Laws*, esp. 738B-C ("in respect of gods, and shrines, and the temples which have to be set up for the various gods in the State, and the gods and daemons they are to be named after, no man of sense, – whether he be framing a new State or re-forming an old one that has been corrupted, – will attempt to alter the advice from Delphi ..."), 759C-D ("They ought to bring from Delphi laws about all matters of religion, and appoint interpreters [ἐξηγηταί] thereof, and make use of those laws") and 828A ("Our next task is, with the help of the Delphic oracles, to arrange and ordain by law the festivals, prescribing what sacrifices, and to what deities, it will be good and right for the State to offer ..."). According to *Republic* 427C, this authority is not only consid-

He is the πάτριος God (the ancestral god) for the Athenians but also the πάτριος ἐξηγητής of all mankind:

> For this God surely is in such matters for all mankind (πᾶσιν ἀνθρώποις) the interpreter of the religion of their father (πάτριος ἐξηγητής) who from his seat in the middle and at the very navel of the earth delivers his interpretation (οὗτος γὰρ δήπου ὁ θεὸς περὶ τὰ τοιαῦτα πᾶσιν ἀνθρώποις πάτριος ἐξηγητής [ἐν μέσῳ] τῆς γῆς ἐπὶ τοῦ ὀμφαλοῦ καθήμενος ἐξηγεῖται). (Plato, *Republic* 427C).

All legislation with regard to "the founding of temples, and sacrifices, and other forms of worship of gods, daemons, and heroes" (427B) resides with Apollo. Hence Plato authorises Socrates's possession of the *daimonion* with the highest possible religious authority, that of Apollo, as the divine πάτριος ἐξηγητής of all mankind.[13]

This way of thinking is closely paralleled in the Gospel of John, according to whom it is Christ who is the ultimate divine exegete as "No one has ever seen God. It is the only-begotten God, who is close to the Father's bosom, who has interpreted him (ἐξηγήσατο)" (John 1:18). Given that at the end of 1 John the author lashes out against pagan gods (1 John 5:21), it seems not unlikely that John is expressing a deliberate criticism of Apollo as the authority in religious matters and replacing Apollo with Christ, the founder of a new, non-locative, universal worship of God "in spirit and truth" (John 4:21–24). Both Socrates and Christ embody or communicate divine signs, and both are at the same time exegetes of what these signs signify.

These various reflections in Plato's writings on the importance of Apollo help to understand why Socrates was closely connected with Apollo. Regardless of what the exact relation is between "Socratic Reason and Socratic Revelation," to use the terminology of the relevant chapter in Mark McPherran's *The Religion of Socrates*,[14] it is clear that there was a close connection between Socrates and

---

ered relevant for the Athenians, but also "for all mankind"; cf. [*Epinomis*] 988A in a reflection about the nobility of Greek religion in comparison with the religion of the barbarians: "... the Greeks (...) have the benefit of their various education, their prophecies from Delphi, and the whole system of worship under their laws." According to *Republic* 540A-C, it is "the Pythian oracle" at Delphi that must approve the deification of the "guardians of the state" after their death, if they have led a philosophical life and taken their turn in fulfilling the offices of the state: "And the state shall establish public memorials and sacrifices for them as to divinities if the Pythian oracle approves or, if not, as to divine and godlike men" (540B-C).

13 For a discussion of the relation between Socrates and Apollo, cf. also Christopher D. C. Reeve, "Socrates the Apollonian?" in *Reason and Religion in Socratic Philosophy*, ed. Nicholas D. Smith and Paul B. Woodruff (Oxford: Oxford University Press, 2000), 24–39.

14 For this discussion, see Thomas C. Brickhouse and Nicholas D. Smith, "Socrates's Gods and the *Daimonion*," in *Reason and Religion in Socratic Philosophy*, ed. Nicholas D. Smith and Paul B. Woodruff (Oxford: Oxford University Press, 2000), 74–88, together with the scholarly corre-

Apollo, and that "the sign of Socrates" is, in all likelihood, the sign given to him by the god Apollo. As regards Socrates's sign, we have seen three things so far: Socrates's sign is apotreptic, clearly divine, and closely related to the god Apollo. As we shall see, the first two features, the apotreptic and divine nature of Socrates's sign, also seem to characterise the signs of Christ in John's Gospel.

7.4 The Signs of Christ: Divine, Divinely Authorised, and Apotreptic

To open the comparison with the divine, Apollonian nature of Socrates's sign, it seems clear that according to John, too, the signs of Jesus are divine and directly related to God. Just as the sign of Socrates is an important feature of Plato's portrayal of Socrates, so the signs of Christ are a very important aspect of John's Gospel. Given the close similarity between John's Gospel and Plato's dialogues, and in particular the dialogues about the last days of Socrates and the *Symposium*, it seems unlikely to me that there existed a separate Signs Source, as Bultmann claimed in his construction of John's Gospel as a multi-layered, multi-authored writing. In my alternative view, the importance which John attaches to the signs of Christ reflects rather the significance of the sign of Socrates in Plato's narrative. I shall now outline how John's depictions of the signs of Jesus help to structure his Gospel, and how this topic of Jesus's signs feeds into, or combines with other issues in the Gospel such as John's portrayal of Jesus's apotreptic behaviour and his use of sympotic settings throughout his gospel.

The first sign (2:11) in John's Gospel, which is consciously portrayed as "the first sign" (2:11), is the wine miracle in the sympotic setting of the wedding at Cana in John 2, where Jesus initially apotreptically declines to assist the ἀρχιτρίκλινος, the president of this reclining wedding symposium (2:8–9), stating that his "hour has not yet come" (2:4). This sign initiates John's portrayal of the signs of Christ, which culminates in the final, or at least most climatic sign: the death and resurrection of Christ himself, already announced in John 2:18–21 and closely related to Jesus's repeated signification of what "kind of death he was to die" in John 12:33 and 18:32. The signs of Jesus help people to come to trust him (2:23), and their significance is recognised by Nicodemus, during his nightly encounter with Jesus which resembles the nocturnal, post-sympotic discourse between Socrates and

---

spondence published in chap. 10, "Socrates and His *Daimonion*," 176–204; and McPherrran, *The Religion of Socrates*, 175–246.

Alcibiades referred to in Plato's *Symposium* (217D–219D), who points out that "no one can do these σημεῖα that you do, unless God is with him" (3:2).

The timing of all these signs reflects an underlying divine chronology, as is also clear from the sign of the healing of the son of the official at Capernaum: this occurrence is carefully enumerated by John as "the second sign that Jesus did when he had come from Judea to Galilee" (4:54; cf. 4:48) and its timing at "the seventh hour" is minutely recorded (4:52–53). The signs of Jesus, however, are not simply magical performances but require an appropriate understanding, as John 6 makes clear. The multitude are attracted to Jesus "because they saw the signs which he did on those who were diseased" (6:2), and they are hosted by Jesus at – to use Kathryn Topper's phrase – "a symposium of the primitive," where they "recline" on the ground (6:11; cf. Mark 6:39) and experience the sign (6:14) of Jesus's multiplication of bread and fish. However, they are subsequently criticised for seeking Jesus not because of the meaning of the signs, but because they labour for "the food which perishes" instead of "the food which endures to eternal life" (6:26–27), and for just demanding another sign (6:30). They fail to understand the significance of the signs as referring to "the true bread" (6:32), and the "true food" (ἀληθής βρῶσις), and the "true drink" (ἀληθής πόσις) (6:55). This criticism of merely physical sustenance and satisfaction, and the exhortation to labour for true, spiritual sustenance is fully paralleled by Socrates's philosophical criticism of "the so-called pleasures" and necessities of eating and drinking in Plato's *Phaedo* (64D, 66B–C; 81B), and his recommendation to embrace "what is really true" (τὸ ἀληθέστατον) and take that as one's only food (65D–E; 84A–B), in this way contrasting "the necessary food (ἡ ἀναγκαῖα τροφή)" (66B–C) with the true, spiritual "food" (τροφή) (84A–B). Jesus's signs, therefore, are not simply magic miracles, but call for a deeper understanding which differentiates, in the same way Socrates does, between a physical reality and a reality which is really true.

According to John 7, Jesus's signs (7:31) continue to draw the attention of many from the crowd during his teaching in Jerusalem at the festival of Booths, where he only arrives at the middle of the feast after his departure for the feast has been apotrepically delayed (7:2–10). The signs subsequently performed in Jerusalem include the healing (9:16) of the rather paradigmatic figure of the man blind from birth, followed by the equally important sign (12:18; cf. 11:47) of the resurrection of Lazarus, again a miracle only performed after Jesus has been apotrepically delayed in his response to the news of Lazarus's illness (11:6). According to the immediate continuation of John 11, following this sign of the resurrection of Lazarus and in view of all the signs performed by Jesus, the Jewish Sanhedrin convenes, raising the question: "What are we to do because this man is performing many signs? (Τί ποιοῦμεν, ὅτι οὗτος ὁ ἄνθρωπος πολλὰ ποιεῖ σημεῖα;) If we let him go on like this, everyone will believe in him, and the Romans will come and destroy both our holy place and our nation" (11:47–48). Now the Sanhedrin formally decides to plan to put Jesus to death (11:49–53), after many less official attempts (5:18; 7:1, 19, 25, 30; 8:37, 40; 11:53) which were unsuccessful because Jesus's hour had not yet come (7:30). Comparing Plato's dialogues about the last days of Socrates and John's Gospel, one could argue that just as Plato portrays the sign of Socrates and his ensuing mission to the Athenians as the cause of the many enmities and widespread hatred against him (*Apology* 22E–23A, 23E–24A), and pictures the Athenian Areopagus as effectively silencing the sign of Socrates by putting him to death – thus running the risk of finding no replacement for him and "passing the rest of [their] lives in slumber, unless God, in his care for [the Athenians], should send someone else" (30A–31A) – , so John depicts the Judean Sanhedrin as planning to silence Jesus because of his signs which they fail to recognise as divine.

When Jesus subsequently comes to Jerusalem, it is during his final teaching in the Jerusalem temple that, according to John 12, he "signifies by what death he was to die (σημαίνων ποίῳ θανάτῳ ἤμελλεν ἀποθνῄσκειν)" (12:33). Jesus leaves behind an audience, some of whom do not believe in him despite the many signs performed in their presence (12:37), and some of whom do believe in him, although not openly because of the fear of suppression (12:42–43). Finally, Jesus is said to repeat his signification of his death at his trial, again "signifying what death he was to die (σημαίνων ποίῳ θανάτῳ ἤμελλεν ἀποθνῄσκειν)" (18:32; cf. 12:33).

The climax of these series of signs consists in the sign of Jesus's death and resurrection (2:18). That the signs of Jesus are regarded as the stepping stones of his public performance is indicated by John's statement at the end of John 20, where the author states that "Jesus did many other signs in the presence of his disciples, which are not written in this book" (20:30), but that those recorded, if properly understood, will enable the readers to acquire eternal life (20:31). Hence, the signs of Christ are as important as the sign of Socrates, which empowers Socrates's philosophical mission among the Athenians. Just as Socrates's sign is the divine sign of Apollo, so equally Christ's signs are the signs of God, as is recognised by Nicodemus: "no one can do these signs that you do if "the god" is not with him (οὐδεὶς γὰρ δύναται ταῦτα τὰ σημεῖα ποιεῖν ἃ σὺ ποιεῖς, ἐὰν μὴ ᾖ ὁ θεὸς μετ' αὐτοῦ)" (John 3:2). Both the sign of Socrates and the signs of Christ are divine and directly related to "the god," whether Apollo or God the divine Father (John 1:12; 4:21, 23; 8:41–42; 20:17).

Both these gods also function as the final justification and authorisation of Socrates's and Jesus's mission. Just as Socrates, in front of the Athenian court, appeals to Apollo himself as the witness and authorisation of his wisdom and ensuing philosophical mission, so Jesus, too, in front of his Jewish opponents, appeals to God as his witness. With regard to Socrates's appeal to Apollo, Plato has Socrates say the following to his judges: "For of my wisdom – if it is wisdom at all – and of its nature, I will offer you the god of Delphi as a witness (μάρτυς)" (*Apology* 20E); he then has him continue with the report of his comrade Chaerephon visiting the oracle of Delphi and receiving confirmation that no one is wiser than Socrates (21A). In this way Socrates's possession of the divine *daimonion* and sign of the god, seen in the charges against him as a sign of his impiety in making new gods, is authorised by Apollo himself. Similarly, Jesus appeals to God's witness in John's Gospel, as a way of authorising his own identity and mission. Against the Jews who persecute and seek to kill him because he makes himself equal with God (5:16–18), Jesus states:

> If I bear witness (μαρτυρῶ) about myself, my evidence (μαρτυρία) is not true. ... The works that the Father has given me to complete, the very works that I am doing, bear witness (μαρτυρεῖ) on my behalf that the Father has sent me. And the Father who sent me has himself borne witness (μεμαρτύρηκεν) on my behalf. (5:31, 36–37)

Moreover, the way this divine confirmation is made accessible to the readers of Plato's *Apology* and John's Gospel is rather similar. Plato has Socrates refer to his old companion Chaerephon, whose question to Apollo's oracle at Delphi elicited the god's response, but because Chaerephon is now dead, Socrates points his judges to Chaerephon's brother who can corroborate the story and bear witness to Apollo acting as witness on behalf of Socrates: "And about these things his brother here will bear you witness (μαρτυρήσει), since Chaerephon is dead" (21A). Plato's authorisation of Socrates by Apollo before the eyes of Plato's readers hence rests on a double process of witnessing: according to Plato, Socrates refers his judges to Chaerephon's brother as the witness to Chaerephon's reception of Apollo's response, in which Apollo acts as a witness on behalf of Socrates. This is fully paralleled by John's authorisation of Christ by God: the author of the Fourth Gospel, who is conventionally called John, is one and the same as the beloved disciple who – together with the other earliest disciples (15:27; cf. 12:17) – bears witness in his Gospel (21:24; cf. 19:35) to Christ's reference to God as his witness (5:37; see also 8:13–14, 17–18 and 10:25; cf. 15:26). Both Plato's *Apology* and John's Gospel thus stipulate rather concrete channels through which the divine testimony about Socrates and Christ, respectively, is mediated.

Not only does John's Gospel reflect Plato's dialogues insofar as the signs of both Socrates and Christ are divine, their recipients divinely authorised, and the knowledge about this mediated through clearly defined human traditions, but, as we have already seen, Jesus's behaviour, like that of Socrates's *daimonion*, is also apotreptic. Both Socrates and Jesus are suddenly delayed in the actions in which they are engaged, or refrain from initiating particular new actions. In John's Gospel this apotreptic portrayal of Jesus is closely connected with the frequent notion that it is not yet the appropriate hour or time to perform a particular action, whether that be a particular miracle, or the ultimate event of Jesus's death. This connection is also present in Plato's dialogue, as Socrates's observation during his trail that the sign of the god no longer apotreptically opposes him (*Apology* 40A–C) is taken as an indication that the "hour" (ὥρα) of his death is approaching (42A); nevertheless, on the morning of the day of his execution (*Phaedo* 59D–60A), "there is still light" for him to try to convince his reluctant, sceptical disciples Simmias and Cebes of the immortality of the soul (*Phaedo* 89C). Both in Plato's writings and John's Gospel the phenomenon of Socrates's and Jesus's apotreptic behaviour seems intrinsically linked with the notion of an underlying divine chronology.

There are also some slight differences. As regards Jesus's apotreptic conduct, we have already briefly seen that some of the signs themselves are also apotreptically delayed, in particular the wine-making miracle in John 2 and the resurrection of Lazarus in John 11. One clear difference from Plato is that in the dialogues the appearance of the sign of the god is the cause of Socrates's delay, whereas in John's Gospel the relation between the signs of Christ and his apotreptic behaviour is more detached. In the case of the miracles of Cana and Lazarus, the performance of the signs themselves is delayed because of Jesus's apotreptic attitude. In other instances the relation is even looser, or is entirely absent.

However, Jesus's unique apotreptic behaviour in John's Gospel, unparalleled in the Synoptic gospels, seems to constitute evidence for John's acquaintance with Socrates's apotreptic sign of the god Apollo as described in Plato's dialogues. Like Plato, John develops his apotreptic portrayal of Christ in a pervasive awareness of a fundamental, all-encompassing divine chronology, which forges all sequences of delays and proper times and hours, of all actions and inactions, into one coherent, divinely-led progression of time. Hence John's explicit timing of particular signs, actions, or occurrences at a specific hour of the twelve hours of daylight (11:9), whether at the sixth (4:6; 19:4), seventh (4:52–53) or tenth (1:39) hour, or at an instantaneous, present hour (4:21, 23; 5:25, 28; 16:32; 19:27), or at an unknown yet destined hour in the future (16:2, 4; 16:21) is an expression of this divine chronology. The frequent Johannine expression that it is not yet the hour (ὥρα) or exact time (καιρός) for a particular miracle (2:4), visit to a festival (7:6, 8), or for Jesus's arrest, trial, and death (7:30; 8:20) is merely the negative expression of this positive view. Two of the most important points on the divine timeline in John's Gospel, and equally constitutive of its narratological unfolding, are of course the final, decisive hour of Jesus's arrest, death, and return from the cosmos to the divine father (7:30; 8:20; 12:23, 27; 13:1; 17:1), and the hour at which

Jesus will no longer speak to his disciples in figures of speech (ἐν παροιμίαις), but plainly (16:25). I will return to this latter timepoint in the final section below.

Before that, however, I will give an integrative overview of John's apotreptic portrayal of Jesus, with reference to Plato where appropriate. In John 2, Jesus declines to help out immediately at the wedding at Cana, stating that his "hour (ὥρα) has not yet come" (2:4), but his mother – apparently aware of Jesus's apotreptic behaviour – already prepares the servants to follow Jesus's instructions (2:5), which he does indeed give at the appropriate time (2:7–8). The next apotreptic situation occurs in John 7, when Jesus's brothers challenge him to reveal himself at the Jewish festival of Booths in Jerusalem (7:2–5). Jesus refuses to go, as his "time (καιρός) has not yet come" (7:6, 8). He stays behind in Galilee until his time has fully come, after which he travels to the festival, only revealing himself "about the middle of the festival," when he starts to teach in the Jerusalem temple (7:10–14). As we have already seen, this profile closely resembles Socrates's delayed arrival at Agathon's dinner party, when they were "about half-way through dinner" in Plato's *Symposium* (174D–175C). Socrates had become "absorbed in his own thoughts" on his way to Agathon's house (174D), just as Jesus had apparently been reflecting on what he was going to teach at the festival. When the Jewish authorities, in response to this teaching, try to arrest Jesus, they are unsuccessful "because his hour (ὥρα) had not yet come" (7:30). Instead, Jesus tells them that he will be with them "a little while longer (Ἔτι χρόνον μικρὸν)" (7:33).

In the case of Socrates's sign, it is a feature of its apotreptic function that it stops Socrates "at many points in the midst of a speech" (*Apology* 40B). This behaviour seems to be mirrored in John's description of Jesus's reflective silence in John 8, when, while he is teaching in the temple, the Jewish scribes and Pharisees interrupt Jesus and ask him whether the woman they have caught in the act of committing adultery needs to be punished in accordance with the Mosaic law (8:1–5). Jesus refrains from answering, absorbed in his thoughts and writing on the ground. Only when they keep on questioning him does Jesus respond, after which he resumes his silence (8:6–8). This kind of silence is absent from the narrative of the Synoptic gospels and seems to reinforce John's apotreptic christology. When Jesus has resumed his teaching in the treasury of the temple, and again refers to the divine Father as his witness (8:12–18; cf. 5:36–37), the Jewish authorities are again unable to arrest him, "because his hour (ὥρα) had not yet come" (8:20; cf. 7:30).

Jesus's apotreptic behaviour comes to the fore yet again in John 11, when he responds to the message of Lazarus's severe illness by deliberately delaying his departure for him by two days (11:1–7; cf. 10:40). When Jesus travels to the feast of Passover in John 12 and is approached by pagan Greeks who have come to worship at the festival, Jesus responds to them in an enigmatic way and refers to the approaching hour of his death (12:20–28). To the crowd which surrounds him, Jesus says: "The light is with you for a little longer (Ἔτι μικρὸν χρόνον τὸ φῶς ἐν ὑμῖν ἐστιν). Walk while you have the light, so that the darkness may not overtake you" (12:35). This way of encouraging the crowd to profit from his presence and teaching while this is still possible is very similar to the way in which Socrates, on the day of his execution, urges his sceptical disciples Simmias and Cebes to let themselves be convinced of his argumentation in favour of the immortality of the soul: "call me to help you …, while there is still light (ἕως ἔτι φῶς ἐστιν)" (*Phaedo* 89C). Just as Socrates offers his help at the final moment before his execution, so Jesus's words to the crowd mark the final stage of his public performance, just before the summary statement of his teaching (12:44–50).

After this conclusion of Jesus's public performance, John states in chapter 13 that at the beginning of the subsequent Last Symposium with his disciples Jesus is fully aware of the fact "that his hour (ὥρα) had come to depart from this world and go to the Father" (13:1). Again, these words seem to resonate with Socrates's final words of defence before the Athenian judges: "But now the time has come to go away (ἀλλὰ γὰρ ἤδη ὥρα ἀπιέναι). I go to die, and you to live; but which of us goes to the better lot, is known to none but God" (*Apology* 42A). This statement reflects the same sovereign, independent, autonomous, and free command of time as Jesus has in John's Gospel. Just as Socrates gets up after his final instructions and goes into another room to bathe so that the women "may not have the trouble of bathing the corpse" (*Phaedo* 116A, 115A), Jesus equally resolutely gets up from the table and washes, cleans, and purifies, not himself, but his disciples (John 13:1–10). And in the same way Jesus, while speaking his final words, already exhorts his disciples to proceed from here: "Rise, let us be on our way" (14:31). It is at this moment of his final words, after the closure of his public ministry, that Jesus, according to John 16, also finds that the hour (ὥρα) has come that he will no longer instruct his disciples "in figures of speech (ἐν παροιμίαις)" but "plainly" (16:25) – an issue to which I will shortly devote the last section of this paper. This moment of final, plain revelation immediately precedes the final hour (ὥρα) which, as Jesus solemnly declares at the beginning of John 17 in his concluding prayer to the divine Father, has now come (17:1).

It does indeed seem that John's apotreptic christology, which is unique in comparison with the christology of the Synoptics, has much in common with Plato's depiction of Socrates's divine apotreptic sign, which delays or suspends his actions.[15] The portrayals of both Socrates's and Christ's apotreptic behaviour

---

[15] It may be worth considering whether Plato's alternative terminology for the "sign" of Socrates (*Apology* 40A–C; 41D) in the dialogues describing the last days of Socrates, i.e. the terminology

are embedded in a divine chronology and are intrinsically connected with what is seen as Socrates's and Christ's divine mission, for which Apollo and God are their

---

of the "*daimonion*" (*Euthyphro* 3B; *Apology* 31C-D; 40A) and the "voice (φωνή)" (*Apology* 31D), is also alluded to in John's Gospel. As regards the terminology of "voice," it seems noteworthy that whereas in the Synoptic gospels the voice of God is restricted to the moments of baptism (Mk 1:11 = Lk 3:22 = Mt 3:17) and the transfiguration on the mountain (Mk 9:7 = Lk 9:35–36 = Mt 17:5), and the voice of Jesus is only emphasised at the moment of his dying (Mk 15:34, 37 = Lk 23:46 = Mt 27:46, 50), in John's Gospel the "voice" of Jesus occurs frequently. References in John's Gospel to the voice of Jesus are found at 3:29 about the voice of the bridegroom; 5:25, 28–29 about the dead who will hear his voice; 10:3–5, 16, 27 about sheep of different origins which will all listen to their shepherd's voice, recognise it, and merge into one flock and follow him; 11:43 about Jesus calling Lazarus from the tomb with a loud voice; and 18:37 about those "who belong to the truth" listening to Jesus's voice; in addition, the voice of God is referred to on two further occasions (5:37 and 12:28, 30). On both occasions there seems to be a subtle interplay with a combination of issues which also occur in Plato's dialogues. The first occasion concerns Jesus's appeal to God as bearing witness to him, an appeal which parallels Socrates's reference to Apollo as his witness (*Apology* 20E). Immediately after Jesus's appeal to God as his witness, John has Jesus state that his opponents have never heard the "voice (φωνή)" of God: "And the Father who sent me has himself borne witness on my behalf. You have never heard his voice (καὶ ὁ πέμψας με πατὴρ ἐκεῖνος μεμαρτύρηκεν περὶ ἐμοῦ. οὔτε φωνὴν αὐτοῦ πώποτε ἀκηκόατε)" (5:37). This statement could be an implicitly polemical remark against competing claims, such as that by Socrates, about having heard a divine voice. The second occasion sees the combination of the notions of the divine *voice* and of Jesus as *signifying* what it means in the present circumstances of his impending death (12:27–33), combining, as it were, the Socratic terminology of the divine voice and the divine sign. There seems to be a similar interaction in John's descriptions of Jesus's voice in John 10. There Jesus's extensive references to his voice and his reflections on its effects and his closeness to God are greeted by his Jewish opponents with the accusation that he has a *daimonion* (10:20), while others contest this (10:21). As regards the terminology of "daimonion," it seems that, compared to the Synoptic gospels, the charge of Jesus having a *daimonion* is more direct and explicit in John's Gospel. In the Synoptic gospels this charge is construed rather differently, as Jesus's opponents feel themselves provoked by Jesus's exorcism and respond by stating that he has Beelzebul, the ruler of the demons, through whom he is able to cast out demons, thus implying that Jesus has a demon (Mark 3:22 = Luke 11:15 = Matthew 9:34 and 12:24). The discourse in John's Gospel is rather different because the notion of casting out demons, which is so prominent in the Synoptic gospels, is entirely absent from it. Moreover, in John's Gospel the charge of Jesus having a *daimonion* is also frequently repeated (7:20; 8:48–49, 52; 10:20–21) and is hence more central to John's concerns: it is levelled against Jesus when he identifies too closely with the divine Father, when he claims that his teaching derives from God (7:14–18), that he has been sent by God (8:38–42), or that he has received his commands from the divine Father (10:14–18). This is more reminiscent of Socrates's claims about his divine mission and his possession of a *daimonion* than of the exorcist background of the Synoptic gospels. Unlike the terminology of "sign" and "voice," the term "daimonion" is impossible for John to apply to Jesus because of its strongly and exclusively negative associations for Jews. Hence Jesus is said to emphatically deny that he possesses a *daimonion* (8:49). It is striking, however, that all the Platonic equivalents of Socrates's *daimonion*, sign, and voice are there in John's Gospel, in exactly the same terminology.

final witness, and which comes to a close when the "hour has come" to depart from the world.

7.5 Apollo's Signification: "Neither Telling nor Concealing, but Signifying"

If Socrates's "sign of the god" is indeed the sign which he has received from Apollo, as I have argued in the third section above, and if John does apply this sign to Jesus, as especially John's apotreptic application and his reference to God as Jesus's ultimate witness indicate, as I have contended in the previous section, the following may also become relevant. In Platonic authors such as Plutarch there are semiotic reflections on Apollo as the signifying god. These reflections take their starting point in an axiom of Heraclitus about Apollo, which reads: "the Lord whose prophetic shrine is at Delphi neither tells nor conceals, but signifies (ὁ ἄναξ, οὗ τὸ μαντεῖόν ἐστι τὸ ἐν Δελφοῖς, οὔτε λέγει οὔτε κρύπτει ἀλλὰ σημαίνει)" (Heraclitus B 93 DK). While Plato himself does not show an acquaintance with this specific Heraclitean axiom, although he does of course know about Heraclitus, the axiom is known to Platonists such as Plutarch and Iamblichus.[16] As we shall see, Plutarch applies this saying from Heraclitus about Apollo as the signifying god in a kind of historiographical way to account for the changes of Apollo's Delphic oracle over time. I will argue that it is useful to analyse Plutarch's semiotic reflections on these changes, as his thought shows a significant correspondence with John's presentation of the changes in Jesus's semiotic strategy. As we have already briefly seen, according to John it is at the end of Jesus's Last Symposium with his disciples, after the conclusion of his public ministry, that Jesus states that the hour (ὥρα) has come that he will no longer instruct his disciples "in figures of speech (ἐν παροιμίαις)" but "plainly" (16:25). As I will argue, this view entails a semiotics which closely resembles Plutarch's view on Apollo's semiotic strategy. I will now first focus on Plutarch's view on the matter, in order to prepare the grounds for a comparison with John.

Plutarch's reflections on Apollo as the signifying god are different from his thoughts about the *daimonion* of Socrates. About the latter issue he reflects in a separate treatise *De genio Socratis* to which I have already referred above, but his views on Apollo's signification are developed in his *De Pythiae oraculis*. In

---

16 See Plutarch, *De Pythiae oraculis* 404D (cf. Plutarch, Fragment 202 [Sandbach]); and Iamblichus, *De mysteriis* 3.15.

this treatise Plutarch comments on the development of the Delphic oracles, which were first given in unintelligible verse, but now in clear prose. It is in the context of this issue that Plutarch refers to the axiom of Heraclitus that "the Lord whose prophetic shrine is at Delphi neither tells nor conceals, but signifies" (ὁ ἄναξ, οὗ τὸ μαντεῖόν ἐστι τὸ ἐν Δελφοῖς, οὔτε λέγει οὔτε κρύπτει ἀλλὰ σημαίνει; B 93 DK)" (*De Pythiae oraculis* 404D–E), and argues that whereas the Delphic oracles used to be expressed in "myths and proverbial figures of speech (παροιμίαι)" (406C), they are now rendered in plain language, leaving behind their past circumlocution, indirectness, and wordiness (406D–F). In religious matters too, according to Plutarch, people have come "to look with suspicion upon metaphors (μεταφοραί), riddles (αἰνίγματα), and ambiguous statements (ἀμφιβολίαι)" (407A–B). Whereas such circumlocution was indispensable in the past, "when there was need of double entendre, indirect statement, and vagueness for the people of ancient days" (407C–D), it is no longer necessary, Plutarch holds, in the present time of Roman "peace and tranquility" (408B). On this understanding, Heraclitus's axiom is taken as a description of the oracular practice of the past, when Apollo's oracular language was still heavily bound up in signs and ambiguities, which neither qualified as full disclosure nor as full concealment,[17] but which have now given way to the age of plain speech and simplicity.

It is exactly this progression from unintelligible παροιμίαι ("proverbial figures of speech") towards full, intelligible disclosure which is reflected in Jesus's progressive semiotic strategy in John's Gospel. The first stage of this strategy comprises Jesus's entire public ministry up to the end of John 12, so up to but excluding Jesus's Last Symposium with his disciples in John 13–17. This first stage is characterised by Heraclitus's statement about Apollo as the signifying god, who "neither tells nor conceals, but signifies" (οὔτε λέγει οὔτε κρύπτει ἀλλὰ σημαίνει; B 93D-K). It is impossible to discern whether John was familiar with this axiom from Heraclitus directly, or had encountered it indirectly through Platonic reflections on the axiom. A direct acquaintance of John with Heraclitus cannot be excluded, given that a Jewish author such as Philo of Alexandria refers to him on a variety of issues, emphasizing his compatibility with, or even his dependence upon Moses (*Quis rerum divinarum heres sit* 213–214; *Legum allegoriae* 1.105–108). Be that as it may, it seems that the tension between this ambiguity of "neither telling nor concealing" runs through the Johannine narrative up to the Last

---

17 On the deliberate ambiguity of Heraclitus's philosophy, see also Myles Burnyeat, *Explorations in Ancient and Modern Philosophy* (Cambridge: Cambridge University Press, 2012), 195–204, "Message from Heraclitus," esp. 200 on Heraclitus B 93 DK.

Symposium of Jesus with the inner circle of his disciples, at which Jesus's semiotic strategy changes.

Until then, however, Jesus's strategy is tuned towards Heraclitus's ambiguous programme of signification between the opposite poles of full disclosure and full concealment. On the one hand, Jesus, despite being the Logos, does not speak clearly, as is reflected in the response of the audience that Jesus's word is a λόγος σκληρός (6:60), a hard word, which is difficult to understand (cf. Critias, fragm. 46 = Pseudo-Aelius Aristides, *Ars rhetorica* 2.2.1.7), and the opposite of a logos which is in accordance with "the simplicity of understanding" (Pseudo-Aelius Aristides, *Ars rhetorica* 2.3.1.6, 2.5.1.2). In this sense, Jesus does not speak (clearly). This position is rather similar to that taken by the author of Fragment 202 in the Sandbach edition of Plutarch's fragments (whether this is indeed Plutarch, or rather Aristoxenus, or another author). According to this author, in his comments on the Pythagorean philosophers, their philosophy is heavily characterised by

> ... its use of symbols, a kind of instruction compounded of speech and of silence, as in a mystic ritual. As a result they do not say: "To those with understanding I shall sing; But close your doors, all ye who are profane", but what they signify is immediately lucid and clear of feature for those to whom it is familiar, but dark and meaningless (ἄσημος) to the ignorant. Just as the Lord who is at Delphi "neither affirms nor conceals but indicates (οὔτε λέγει οὔτε κρύπτει ἀλλὰ σημαίνει)", to quote Heraclitus, so with the Pythagorean symbols what seems to be made known is really being concealed (καὶ τὸ φράζεσθαι δοκοῦν κρυπτόμενόν ἐστι), and what seems to be concealed is discerned by the mind (καὶ τὸ κρύπτεσθαι νοούμενον). (Plutarch, *Fragm.* 202, ed. Sandbach)

This aptly captures the ambiguity of Jesus's public performance in John's Gospel, whose words are obviously "hard," "dark and meaningless to the ignorant."

On the other hand, however, Jesus did not completely conceal (himself) either. He did so occasionally, as two passages make clear which speak of Jesus hiding himself from his opponents (John 8:59 Ἰησοῦς δὲ ἐκρύβη; 12:36 καὶ ἀπελθὼν ἐκρύβη ἀπ' αὐτῶν); but these exceptions only draw attention to the fact that, although he did not speak, he did not conceal either. And when questioned by the high priest at his trial about his disciples and his teaching, Jesus refers to his public teaching: "I have spoken openly to the world; I have always taught in synagogues and in the temple, where all the Jews come together. I have said nothing in secret (καὶ ἐν κρυπτῷ ἐλάλησα οὐδέν)" (18:19–20).

The Johannine Jesus, then, perfectly embodies the ambiguity of Heraclitus's Apollo, as he "neither tells nor conceals." This ambiguity, however, only characterises Jesus's public ministry. After he has finished his public performance in John 12, and finds himself in the seclusion of the inner circle – especially once the only "unclean" disciple, Judas, has left, and the others have been symbolically purified (13:5–11; 15:3), instructed, and consoled in the course of John 13–16 – it is at the very end of John 16, just before Jesus's dedicatory prayer on behalf of his disciples in John 17, that the second, climactic phase of Jesus's semiotic strategy commences. Then Jesus tells his disciples:

> I have said these things to you in παροιμίαι, in figures of speech (Ταῦτα ἐν παροιμίαις λελάληκα ὑμῖν). The hour is coming when I will no longer speak to you in figures, but will tell you plainly of the Father (ἔρχεται ὥρα ὅτε οὐκέτι ἐν παροιμίαις λαλήσω ὑμῖν ἀλλὰ παρρησίᾳ) … I came from the Father and have come into the world; again, I am leaving the world and am going to the Father. (John 16:25–28)

These words are now welcomed, in the disciples' immediate response, as direct, clear speech:

> Yes, now you are speaking plainly, not in any figure of speech (Ἴδε νῦν ἐν παρρησίᾳ λαλεῖς, καὶ παροιμίαν οὐδεμίαν λέγεις)! Now we know that you know all things, and do not need to have anyone question you; by this we believe that you came from God. (16:29–30)

No longer does Jesus speak in the figures of speech which characterised, for instance, his metaphorical speech about the sheep that recognise the good shepherd's voice – a statement that John explicitly classifies as a παροιμία, a figure of speech (10:6).

This development from figures of speech to plain speech is exactly the same development as we have detected in Plutarch's *De Pythiae oraculis*. Put in Plutarch's remarkably similar language, the "myths and proverbial figures of speech (παροιμίαι)," with all their circumlocution, indirectness and wordiness, have given way to plain language. And just as according to Plutarch such circumlocution was necessary in the past, "when there was need of double entendre, indirect statement, and vagueness for the people of ancient days" (407C–D), but is now no longer needed under the aegis of Roman imperial "peace and tranquility" (408B), John, too, takes a similar perspective. Whereas there was need for hard speech and figures of speech during Jesus's public performance, there no longer is at the very moment of his return to the Father; now for the first time in John's Gospel Jesus's peace is proclaimed, during the Last Symposium, to the disciples (14:27, 16:33; cf. the post-resurrection peace in 20:19, 21, 26).

It is tempting to state that for John not only Jesus's "hard speech (λόγος σκληρός)" and "figures of speech (παροιμίαι)" have been supplanted by his plain and simple words, but that also the "signs" (σημεῖα) themselves have been superseded. Although John, as we have seen, models his apotreptic christology on Socrates's apotreptic sign of the god Apollo, it does indeed seem that the signs lose importance in John's narrative after Jesus's full and final disclosure of his mission to his disciples. Although in formal terms Jesus's death and resurrection are referred to as a sign during Jesus's public performance in John 2:18 and 12:33, the sign language recedes after the section of the Last Symposium and its revelation of Jesus's mission. This language recedes, only to re-emerge in a brief authorial reflection on Jesus's response to Pilate (18:32; cf. the same wording in 12:33) and in the Gospel's postscript to the readers, in which John states that "Jesus did many other signs in the presence of his disciples, which are not written in this book," but that those contained in his writing enable the readers to obtain eternal life (20:30–31). These two passages after the section on the Last Symposium concern either Jesus's words to non-disciples such as Pilate, spoken in public, or the author's words to the general readers, who may not yet have encountered the true and full meaning of Jesus's message. For his inner circle of "purified" disciples (13:10–11; 15:3), however, assembled at the Last Symposium, Jesus has abandoned his hard speech and figures of speech, and also his signs, because he now speaks plainly (16:25–30). Hence John's narrative can be understood as a narrative which is largely embroidered on the Heraclitean axiom of God "neither telling nor concealing, but signifying," while clearly marking the essential transition from the indirect, diffuse circumlocution of λόγοι σκληροί, παροιμίαι and σημεῖα to plain speech at the crucial moment of the Last Symposium.[18] Here religious knowledge is no longer hidden but fully shared, accessible, without hindrance, for all readers of John's Gospel.

References

Myles Burnyeat, *Explorations in Ancient and Modern Philosophy* (Cambridge: Cambridge University Press, 2012).

Rainer Hirsch-Luipold, "Klartext in Bildern: ἀληθινός κτλ., παροιμία – παρρησία – σημεῖον als Signalwörter für eine bildhafte Darstellungsform im Johannesevangelium," in *Imagery in the Gospel of John: Terms, Forms, Themes and Theology of Johannine Figurative Language*, ed. Jörg Frey, Jan G. van der Watt, and Ruben Zimmermann, Wissenschaftliche Untersuchungen Zum Neuen Testament 200 (Tübingen: Mohr Siebeck, 2006), 61–102.

George van Kooten, "The Last Days of Socrates and Christ: *Euthyphro, Apology, Crito*, and *Phaedo* Read in Counterpoint with John's Gospel," in *Religio-Philosophical Discourses in the Mediterranean World: From Plato, through Jesus, to Late Antiquity*, ed. Anders K. Petersen and George van Kooten, Ancient Philosophy & Religion 1 (Leiden: Brill, 2017), 219–43.

---

18 Cf. also Rainer Hirsch-Luipold, "Klartext in Bildern: ἀληθινός κτλ., παροιμία – παρρησία – σημεῖον als Signalwörter für eine bildhafte Darstellungsform im Johannesevangelium," in *Imagery in the Gospel of John: Terms, Forms, Themes and Theology of Johannine Figurative Language*, ed. Jörg Frey, Jan G. van der Watt, and Ruben Zimmermann, Wissenschaftliche Untersuchungen Zum Neuen Testament 200 (Tübingen: Mohr Siebeck, 2006), 61–102.

George van Kooten, "The 'True Light which Enlightens Everyone' (John 1:9): John, Genesis, the Platonic Notion of the 'True, Noetic Light,' and the Allegory of the Cave in Plato's *Republic*," in *The Creation of Heaven and Earth: Re-interpretations of Genesis I in the Context of Judaism, Ancient Philosophy, Christianity, and Modern Physics*, ed. George van Kooten, Themes in Biblical Narrative 8 (Leiden: Brill, 2005), 149–94.

Anthony A. Long, "daimōn," in *The Continuum Companion to Plato*, ed. Gerald A. Press (London: Continuum, 2012), 152–54.

Christopher D. C. Reeve, "Socrates the Apollonian?" in *Reason and Religion in Socratic Philosophy*, ed. Nicholas D. Smith and Paul B. Woodruff (Oxford: Oxford University Press, 2000), 24–39.

Kathryn Topper, *The Imagery of the Athenian Symposium* (Cambridge: Cambridge University Press, 2012).

Clare Wilde

8 "They Wish to Extinguish the Light of God with Their Mouths" (Qur'ān 9:32): A Qur'ānic Critique of Late Antique Scholasticism?

8.1 Introduction

The Qur'ān contains a number of references to knowledge and the modes of its transmission. For example, in addition to *kitāb* (book), the Qur'ān has numerous allusions to writing media, such as *asfār/sifr* (book/volume); *khātam* (seal – of the prophets); *lawḥ* (board/tablet); *midād* (ink); *nuskha* (copy/exemplar: Qur'ān 7:154 – Moses' tablets); *qalam* (pen – made of reed; also tubes); *qirṭās/qarāṭīs* (parchment/papyrus: Qur'ān 6:7, 91); *raqq* (parchment: Qur'ān 52:3); *sijjil* (parchment scroll – in an apocalyptic context); *ṣuḥuf* (pages of scripture).[1] Knowledge is linked with faith – as something God has given prophets.[2] It is also one of the attributes of God (as "knower of the seen and unseen": *'ālim al-ghayb wa-l-shahāda*; e.g. Qur'ān 59:22). Additionally, the Qur'ān references the learning/knowledge found in earlier communities – especially relating to the knowledge of God's revelations. Sometimes these references are positive (or at least neutral), as in Qur'ān 10:94's exhortation that, if in doubt about the revelation, he (e.g., Muḥammad) should "ask those who were reading the book before" (cf. also Qur'ān 26:195–197, in which the truth of the Arabic revelation is said to be in the earlier books and known to the *'ulamā'* – scholars[3] – of the Children of Israel).

---

[1] For further discussion and bibliography, see Jane D. McAuliffe, ed., *The Encyclopaedia of the Qur'ān* (Leiden: Brill, 2001–2006), s.v. "Writing and Writing Materials" (a fine piece by Sheila Blain). Hereafter this resource is abbreviated as *EQ*. The bibliography to this article contains full bibliographical details of the *EQ* articles.
[2] For further discussion and bibliography, see *EQ*, s.v. "Knowledge and Learning."
[3] *'Ulamā'* ("scholars") twice appear in the Qur'ān (26:197 and 35:28). Additionally, those "firmly grounded in knowledge" of Qur'ān 4:162 have occasionally been glossed as the scholars of the Children of Israel; see Louise Marlow's *EQ* article "Scholar."

Note: I first approached this topic at the encouragement of Fr. David Johnson, as a seminar paper at CUA. A very early version was presented at the IV International Syriac Symposium in Princeton, NJ. On the occasion of Prof Angel Urban's Festschrift, it was later reworked. Prior to the April 2015 'Sharing and Hiding Religious Knowledge' conference at Groningen, yet another iteration was delivered at the SBL as part of the first IQSA meeting in Baltimore in November 2013. My thanks to all who have encouraged, and contributed to, my thinking on this topic. All errors are my own.

8.2 Distorting God's Word

But, this knowledge (of God's books) – and a (scholarly?) cover-up of what appears in them also appears in qur'ānic polemics, generally directed at the Children of Israel – Jews, but also Christians.[4] This tension between qur'ānic references to the (divine) book and human manipulations (be it of the *ma'āna*, sense, or the *lafẓ*, text[5]), whether oral or written, as well as the rationale (and intentionality) of such distortion, has been the subject of Muslim exegesis and western scholarship on the Qur'ān.[6]

Although the root letters *ḥ-r-f* are the traditional focus of scholars examining the qur'ānic charge of the scriptural corruption by earlier communities, qur'ānic passages that do not contain words derived from these root letters have also been adduced as proof of qur'ānic awareness of Jewish alteration of the biblical text. For, from its allusions to both written and oral manipulations of God's word, a qur'ānic awareness of rabbinic traditions has been posited.[7] For example, might Qur'ān 2:79's allusion to "those who write the book with their hands and then say: This is from God, so that they may take for it a small price ..." reflect qur'ānic awareness of collections of midrashim that already existed in the written form in the seventh century? Did it know of targumim that were traditionally transmitted in the oral form, as, at Qur'ān 3:78, it criticizes "a party of them who distort the Scripture with their tongues, that you may think that what they say is from the Scripture, when it is not from the Scripture. And they say: It is from God when it is not from God; and they speak a lie concerning God knowingly." Alternatively,

---

[4] The verb *ḥarafa* (to distort) is used in the Qur'ān referencing people who hear or listen (*s-m-'* 2:75) to the *kalām allāh* (word of God) then *ḥarrafa* it – intentionally (*'-q-l*) after they learned/ knew it (*'-l-m*) (cf. Qur'ān 4:46; 5:13; 5:41; 2:75). For the still useful classic discussion of this topic, see Ignazio di Matteo, "Il 'tahrif' od alterazione della Bibbia secondo i musulmani," *Bessarione* 38 (1922): 64–111, 223–60.
[5] For a helpful recent discussion of this distinction see Gabriel S. Reynolds, "On the Qur'anic Accusation of Scriptural Falsification (taḥrīf) and Christian Anti-Jewish Polemic," *Journal of the American Oriental Society* 130 (2010): 189–202.
[6] See, for example, the substantive study of Camilla Adang, *Muslim Writers on Judaism and the Hebrew Bible: From Ibn Rabban to Ibn Hazm* (Leiden: Brill, 1996), in which the works of ten Muslim authors are examined. She demonstrates that only three of these ten, the later authors, understood the text of the Hebrew Bible to have been distorted. More recently, Arye Olman proffers a thoughtful and thought-provoking reading of the qur'ānic charge in Qur'ān 6:91 as a response to the Jewish practice of *midrash*. See his "'The Jews Distorted Torah': An Attempt to the Moslem Claim," in *Proceedings of 15th Conference of Jewish Studies*, FSU 'Sefer' 2 (2008): 90–100 (also available on academia.edu). On the qur'ānic charge of Jewish oral distortion of scripture, see esp. Reuven Firestone, "The Failure of a Jewish Program of Public Satire in the Squares of Medina," *Judaism* 46 (1997): 439–52.
[7] The following examples are taken from Olman, "'The Jews Distorted Torah.'"

Qur'ān 6:91's reference to the "Book that Moses brought, a light and a guidance to men, which you make into scattered writings which you show while you conceal much? And you were taught what you did not know, (neither) you nor your fathers" resonates with the Hebrew tradition of the Mishnaic era that had some rules about not reading certain excerpts from Torah in synagogue to avoid their objectionable understanding:

> The story of Reuben is read but not explained; the episode of Tamar is read and interpreted; the first story of the Calf is read and translated, and the second account is read but not interpreted; the Priestly benediction and the narrative of David and that of Amnon are neither read nor translated. (Mishnah Megillah 4:10)[8]

But, the Qur'ān has not been understood as criticizing Jews alone for such scriptural distortions – and, although Islamic tradition often interpreted allusions to scriptural distortion as references to the People of the Book's (*ahl al-kitāb*) willful concealment of mentions of Muḥammad in the Torah, for example, it could also understand these passages as criticisms of Jewish (or Christian) violations of the laws of God, or of Christian (or Jewish) misrepresentations of God as having a Son (e.g., Jesus or Uzayr/Ezra). Exegetes would ask who, exactly, was distorting scripture with their tongues, what was the content of that which they were concealing, what is the exact meaning of *rabbānīn*, how should the studying/teaching verbs be vocalized, what is meant by *al-kitāb* (book: the Qur'ān? The scripture of Jews/Christians?), was the verse revealed in response to questions from Jews and Christians to Muḥammad: "Are you asking us to take you as a lord?," or, are the verses alluding to Jewish and/or Christian mishandling of the scriptures (as when Christians say that Jesus is the Son of God, or when Jews say that Ezra is).[9]

These exegetical discussions point to the multiple understandings possible for these qur'ānic passages. For, often the verbal subject of the phrases indicating the corruption of God's revelation is indeterminate (third person masculine plural verbs or pronouns are frequent) and later exegetes are not unanimous in their interpretations of the verses in question. Furthermore, many of the qur'ānic discussions of the "Children of Israel" and "People of the Book" are ambiguous, and there is no scholarly consensus as to the exact identities of who is intended by these phrases,[10] nor by the explicit mentions of "Jews" (*yahūd*) and "Christians"

---

8 Cited by Olman, "'The Jews Distorted Torah.'"
9 Although the details of the rich and varied interpretation of Qur'ān 3:75–80 preserved in classical works of *tafsīr* are beyond the scope of this paper, these questions are found in numerous *tafāsīr* ad Q 3:75. See, e.g., the comments of al-Tabari and al-Razi, ad loc.
10 For a solid introduction (with relevant bibliography) to this complex issue, see the *EQ* articles "People of the Book," "Jews and Judaism," "Christians and Christianity," "Children of Israel."

(*naṣārā*, as well as "People of the Gospel").[11] Additionally, terms that appear to designate particular functions (e.g. scholars, rabbis, priests, monks) are not uniformly glossed by the classical exegetes. (*Aḥbār*, for example, is generally glossed as rabbis – but can also be understood as Christian scholars.)

Given the variety of later exegetical glosses of *aḥbār* and the multivalent qur'ānic estimations of monks/monasticism (*ruhbān/rahbāniyya*),[12] might we be able to read the qur'ānic discourse on the oral destruction – whether with tongues (e.g., Qur'ān 3:78 or 4:46) or mouths (e.g., Qur'ān 24:15) of God's word as indicative of a familiarity with Late Antique debates (especially in Syriac) over Hellenic education practices by both Jews and Christians?[13]

8.3 Qur'ān 9:30–34: An Allusion to Late Antique Scriptural Study Debates?

The qur'ānic allusions to people (especially Jews) using their tongues to distort Scripture – by adding words (Qur'ān 3:78), or by moving words around (e.g., Qur'ān 4:46) – have been studied in the context of discussions of "*taḥrīf*," the qur'ānic charge that the earlier revelations (to Moses, Jesus, etc.) have been distorted by members of the earlier monotheistic communities. The following argues

---

11 See, e.g. François De Blois, "Naṣrānī (Ναζωραῖος) and Ḥanīf (ἐθνικός): Studies on the Religious Vocabulary of Christianity and of Islam," *Bulletin of the School of Oriental and African Studies* (2002): 1–30 and Sidney H. Griffith, "Al-Nasara in the Qur'an: A Hermeneutical Reflection," in *New Perspectives on the Qur'ān: The Qur'ān in Its Historical Context 2*, ed. Gabriel S. Reynolds (New York: Routledge: 2012), 1–38.
12 For a recent reading of the qur'ānic *rahbāniyya* see Emran I. El-Badawi, "From 'Clergy' to 'Celibacy': The Development of Rahbaniyya between the Qur'an Hadith and Church Canon," *Al-Bayan: Journal of Qur'ān and Ḥadīth Studies* 11/1 (2013): 1–14.
13 On the significance of Syriac for qur'ānic studies see, e.g. Sidney Griffith, "Syriacisms in the Arabic Qur'ān: Who Were "Those Who Said 'Allāh is Third of Three'" According to al-*Mā'idah* 73?" in *A Word Fitly Spoken: Studies in Mediaeval Exegesis of the Hebrew Bible and the Qur'ān; Presented to Haggai Ben-Shammai*, ed. M. Bar-Asher et al. (Jerusalem: Hebrew University Press, 2007), 83–110. Two other recent studies highlight the complexity of attempting to understand the Qur'ān in its Late Antique environment: Emran I. El-Badawi, *The Qur'an and the Aramaic Gospel Traditions* (New York: Routledge, 2013) and Holger Michael Zellentin, *The Qur'ān's Legal Culture: The Didascalia Apostolorum as a Point of Departure* (Tübingen: Mohr Siebeck 2013). Angelika Neuwirth, "Qur'ān and History—A Disputed Relationship: Some Reflections on Qur'ānic History and History in the Qur'ān," *Journal of Qur'anic Studies* 5/1 (2003): 1–18, provides a comprehensive and accessible overview of some of the challenges of studying the Qur'ān in the context of Late Antiquity.

that yet another qur'ānic allusion to the oral destruction of God's word (Qur'ān 9:30–34) might also shed light on qur'ānic engagement with Late Antique debates over the propriety of probing with one's intellect into the meaning of scripture. This passage reads as follows:

- 9:30 And the Jews say: Ezra is the son of Allah; and the Christians say: The Messiah is the son of Allah. These are the words of their mouths. They imitate the saying of those who disbelieved before. Allah's curse be on them! How they are turned away!
- 9:31 They take their *aḥbār* and *ruhbān* (doctors of law and their monks) for Lords besides Allah, and (also) the Messiah, son of Mary, And they were enjoined that they should serve one God only – there is no god but He. Be He glorified from what they set up (with Him)!
- 9:32 They desire to put out the light of Allah with their mouths, and Allah will allow nothing save the perfection of His light, though the disbelievers are averse.
- 9:33 He it is Who sent His Messenger with guidance and the Religion of Truth, that He may cause it to prevail over all religions, though the polytheists are averse.
- 9:34 O you who believe, surely many of the *aḥbār* and *ruhbān* (doctors of law and the monks) eat away the property of men falsely, and hinder (them) from Allah's way. And those who hoard up gold and silver and spend it not in Allah's way – announce to them a painful chastisement[14]

A number of features bear mentioning. First, it does not contain the root *ḥ-r-f*, so is not generally included in scholarly discussions of the qur'ānic charge that Jews and Christians have "distorted" their original revelations. Secondly, it is "mouths" rather than "tongues" that are the means of destruction. Thirdly, the passage contains an explicit mention of "Jews" and "Christians": it is with their mouths that Jews claim Uzayr is the Son of God and Christians, that the Messiah is (Qur'ān 9:30). Fourthly, in this passage, twice *aḥbār* and *ruhbān* are paired: Jews and Christians are accused of taking them (and Jesus) as lords besides God. And, a few verses later, these *aḥbār* and *ruhbān* are accused of taking the wealth of the people, and hoarding gold and silver – and not spending it in the path of God ... all the while preaching about the dreadful torment (of hell). It is in the context of this passage that the charge that "they wish to extinguish the light of God with their mouths" occurs. The antecedent for the "they" of "they wish"

---

[14] Translation found at http://www.aaiil.org/text/hq/englishholyqurantranslation/english-holyqurantranslationmaulanamuhammadali.shtml

(*yurīdūn*, Qurʾān 9:32) is not specified, and exegetes proffer various explanations.[15] Finally, it should be noted that "scripture" is not explicitly mentioned in the passage: rather, people are accused of wanting to use their mouths to "destroy the light of God."

8.4 Christian and Jewish Authorities in the Qurʾān

Classical exegetes and contemporary scholars have explored the Qurʾān's ambiguous estimations of Jews and Judaism and Christians and Christianity (including monasticism) with a number of hypotheses: e.g., various encounters between Muhammad and different Jewish and Christian communities; indications of lingering remnants of a Jewish-Christian sect[16]; polemics.[17] Islamic tradition has discussed the significance of qurʾānic criticisms of Jewish religious authorities (Qurʾān 2:75; 5:44, 63; 9:31, 34),[18] and the similarities of the Jewish and Christian abuses to which the Qurʾān alludes – often in terms of rabbis and priests as having served as lawmakers for their respective communities, sometimes "allowing what God forbade" or "forbidding what God allowed." And, in their obedience to the "new" laws, the communities were effectively abandoning the worship of God and following the religious authorities as their new lords.[19]

---

15 See, e.g., the glosses provided by Fakhr al-Dīn al-Rāzī, Muqātil b. Sulaymān and al-Ṭabarī ad Qurʾān 9:32.
16 Recently, Zellentin, *Legal culture*.
17 E.g. Griffith, "Syriacisms."
18 The qurʾānic critique of rabbis in the aforementioned passages certainly merits further scholarly investigation. Does it, for example, reflect actual developments in the Jewish community of Late Antique Arabia? (cf. Daniel Boyarin, *Border Lines: The Partition of Judaeo-Christianity* [Philadelphia: University of Pennsylvania Press, 2004]). Is it an example of qurʾānic employment of Christian *Adversus Judaeos* argumentation? For some scholarship on this literature, see, e.g., the references in Aryeh Kofsky, "Eusebius of Caesarea and the Christian-Jewish Polemic," in *Contra Iudaeos: Ancient Polemics between Christians and Jews*, ed. Ora Limor and Guy G. Stroumsa, Texts and Studies in Medieval and Early Modern Judaism 10 (Tubingen: Mohr Siebeck, 1996), 59–84, at 65 n. 19. Or is it an indication of qurʾānic awareness of Jewish-Christian polemics—akin to its assertion at Qurʾān 3:67 that Abraham was neither a Jew nor a Christian, but a Muslim, a *ḥanīf*?). This line of investigation is, however, beyond the scope of the present discussion.
19 Cf. e.g. the *tafāsīr* of al-Ṭabarī and al-Rāzī, ad loc. An online version of *Tafsir* Ibn Kathir states that "fighting the Jews and Christians is legislated because They are Idolators and Disbelievers." A more nuanced rendering of classical *tafsir* of Qurʾān 9:31 is found at http://islamicsystem.blogspot.co.nz/2012/06/tafsir-of-surat-al-tawba31-taking.html

Elsewhere (e.g. Qur'ān 5:63), the *aḥbār* are paired with *rābāniyūn* – commonly interpreted as Jewish learned men who should be forbidding people from uttering and eating unlawful things. In yet another passage, the *ruhbān* are paired with *qissīsīn* (priests, Qur'ān 5:82) – but in extremely favorable terms, and in marked contrast to Jews and polytheists. In Qur'ān 9:30 f., however, *aḥbār* and *ruhbān* are painted with the same brush. In classical exegesis, this coupling in Qur'ān 9:30–34 has been examined in the context of narratives about events in the life of Muhammad. And, while Islamic tradition is clear that rabbis and monks are not literally worshiped by Jews and Christians, they were heeded in place of God, insomuch as they served as the "lawgivers" for their respective communities – and made lawful that which God forbad, and vice versa. But, might the coupling of *aḥbār* and *ruhbān*, in the context of the charge of their oral distortion of God's word, also indicate a qur'ānic familiarity with the debates over the "scholasticism" of late antique rabbinic and Christian schools?

8.5 Qur'ānic Critique of Late Antique Scholasticism?

If understood as referencing the practices of Jewish and Christian scholars, the qur'ānic linkage of *aḥbār* and *ruhbān* and their association with an "oral" destruction of God's word (Qur'ān 9:30–34) evokes the scholastic movement among both Jews and Christians in Late Antique Mesopotamia: in Nisibis at the end of the fifth century, and in Seleucia itself in the middle of the sixth.[20] As Adam Becker has discussed,[21] these schools developed in the same time period and place as did the Babylonian Rabbinic academies. These schools also figure in Late Antique eastern Syriac monastic discourse. For example, in response to the Catholicos Iso'yahb's (d. 659) restoration of the monastery at which he had studied, which included plans to "build a school (Syr. *eskole*) in the place of his cell, ...so that a monastery of instruction ... might be accessible to every 'school student,' trained and illuminated in the scripture, so that the school and monastery might become one ...,"[22] a group of monks, including the head of the monastery replied:

---

[20] See Adam Becker, "The Comparative Study of 'Scholasticism' in Late Antique Mesopotamia: Rabbis and East Syrians," *Association of Jewish Studies Review* 34 (2010): 91–113. The discussion of the establishment of these schools is found at p. 94.
[21] Adam H. Becker, *Fear of God and the Beginning of Wisdom: The School of Nisibis and the Development of Scholastic Culture in Late Antique Mesopotamia* (Philadelphia: University of Pennsylvania Press, 2006).
[22] Becker, *Fear of God*, 169.

> This work is not one that belongs to ascetics ... we who sit in our cells. The songs of the hallelujahs, the psalms, the responses, the harmonies of the youths and the vigilant ... will vex us. For we did not find it in a book nor did we receive from report (the tradition) that this thing (i.e., a school) was in one of the monasteries of the fathers. Rather, we ourselves are summoned to weeping and mourning while sitting in our cells ..., according to the teaching which is from scripture and which we have also received from our father Mār Jacob. For he did not order us in his life and in his migration from us that one should teach the other chanting or how to read a manuscript. Leave off making us "school students" again, rather, (let us be) while we sit in our cells ... and (there may be) the solitary reading (of scripture) of each person by himself.[23]

Although this account is found in a ninth-century work, it reflects the tensions within east Syrian monasticism since at least the mid-sixth century. Many east Syrian solitary monastics would have been trained as youths in the schools, but left the school for the monastery and, eventually, the solitude of the desert. Some, however, would have stayed within the school environment, continuing their training and becoming teachers themselves. But, as Becker argues, "group study at the school ... would have served as a devotional practice that bore as much religious significance as prayer and private reading did in institutions focused more on private inspiration, such as those East-Syrian monasteries where monks focused on the higher levels of contemplation as advocated in the writings of Evagrius of Pontus."[24] From the sixth century and their translation into Syriac, the works of the fifth-century Evagrius would prove highly influential in east Syrian monastic spirituality. For Evagrius, prayer is "communion of the mind with God" – a communion that ought to have no intermediary, be they images or even the human voice:

> Every proposition has a genus, which is predicated, or a difference, or a species, or a property, or an accident or what is compounded of these: but nothing which is said in regard to the Holy Trinity is acceptable. Let the ineffable be worshiped in silence.[25]

The dangers of the scholastic methods in scriptural investigation are outlined by the seventh century Dadisho of Bayt Qatraya (a region identified as modern-day Bahrain and eastern Arabia, i.e., Qatar), who "makes a number of passing jabs and disdainful references to *eskolāyē*, that is, school men and students" providing "explicit information on the tension that could exist between the school and the monastery."[26] Dadisho also identifies the "alternative means" the demons

---

23 Quotation found in Becker, *Fear of God*, 170.
24 Becker, *Fear of God*, 172.
25 Quoted in Becker, *Fear of God*, 177.
26 Becker, *Fear of God*, 188.

have devised to trick men: "constant and disordered meditation on scripture, a disparate wandering after seeking its meanings, and suspension of labors, of prayer, of reflection on God, and of meditation and self-correction," – all of which "leads Christians to engage in intellectual disputes."[27] Such criticism, however, did not appear only with Dadisho, or even Evagrius; in fact, this echoes earlier Syriac reflections on the traps into which men fell when they attempted to probe, with their intellect, into the meanings of the names of God, rather than engaging in meditative contemplation on the wonders of God. For example, in a panegyric against Arius (b. 260 CE) concerning theological inquiry, speculation and investigation into the nature of God, the fourth century Ephraem[28] writes[29]:

> Take life from the greatness (of God)
> And leave aside investigation into the greatness.
> Love the grace of the father
> And do not probe into his being.
> Take delight in and love the goodness of the son
> And do not probe into his birth.
> Love the descent of the holy spirit
> And do not apply yourself to investigating it.
> Father and Son and Holy Spirit
> By their names they are understood.
> Do not ponder their hypostases.
> Meditate on their names.
> If you inquire into the being you are brought to naught
> But if you believe in the name you live…

Similarly, Dadisho defends "spiritual" (monastic allegorical) exegesis in the hope that he might

> muzzle the mouths of certain stupid exegetes who, thanks to their knowledge of the jargon that they have learnt – jargon that is totally divorced from any idea of good conduct – hold saints in contempt when these latter introduce examples from the Scriptures and from the natural world, and take them to refer to godliness and righteousness.[30]

---

27 Becker, *Fear of God*, 189.
28 See Sidney H. Griffith, "Images of Ephraem. The Syrian Holy Man and his Church," *Traditio: Studies in Ancient and Medieval History, Thought and Religion* 45 (1989–90): 7–33, for an overview of the man, his church and his portrayal in later literature.
29 *On Faith* (no. 4), lines 121–34, in Edmund Beck, ed., *Des heiligen Ephraem des Syrers Sermones de Fide*, Corpus Scriptorum Christianorum Orientalium 212 (Leuven: Peeters, 1961), 32–36.
30 Quoted by Becker, *Fear of God*, 189 n. 133.

Although the third person (masculine) plural antecedent of those who wish (*yurīdūn*) to extinguish God's light with their mouths in Qurʾān 9:32 is not explicitly identified, the accusation occurs directly after accusations against Jews and Christians for saying, respectively, that Uzayr and the Messiah are sons of God – and the assertion that *ruhbān* and *aḥbār* had been taken as lords besides God. Might the Qurʾān's first auditors, then, have heard Qurʾān 9:32 – and other qurʾānic allusions to the use of tongues and mouths (as well as writing) in the distortion of scripture (e.g., Qurʾān 3:75f.) – as reflecting Late Antique criticism of the scholasticism in which certain contemporary Christians, as well as Jews, engaged?

If its first auditors did indeed hear this passage as directed at monks, as well as at scholastic (?) authorities, both for wishing to extinguish the "light" of God – with their mouths – and for hoarding wealth, such a reading coincides with Becker's reading of criticisms of the pedagogy of the school of Nisibis – that, rather than being distinct from the monastic movement, the scholastic movement was an integral, if contested, part of east Syrian monasticism, indicative of increasing Hellenization, a process that had not just linguistic but also academic and, arguably, administrative, effects.[31] This possibility is particularly convincing in the light of the early exegetical gloss provided by Muqātil b. Sulaymān of the *aḥbār* of Qurʾān 9:32 as both Christian and Jewish *'ulamāʾ* (scholars). In this reading, the qurʾānic criticism of those who destroy the light of God with their mouths is directed at Christian and Jewish scholars, reflecting, perhaps their shared "pedagogical paradigm."[32] If the Arabic Qurʾān is in fact criticizing some contemporary Jewish and Christian scholars, and monks, as being similarly trained, in a classical Hellenic pedagogy, does this indicate an awareness familiarity with, and criticism of, Hellenizing trends? Particularly if the qurʾānic message is understood as located in, and directed at, people familiar with the solitude of the desert, it is not impossible that it would expect its auditors to be familiar with an Evagrian spirituality that would prioritize silent meditation on, rather than scholastic debates about, scripture.

8.6 Conclusion

The Qurʾān presumes its auditors' familiarity with both Judaism and Christianity. But – as indicated by the difference of scholarly opinions about the meaning of qurʾānic verses, who, exactly, were the Jews and Christians the Qurʾān presumes

---

31 See esp. Becker, *Fear of God*, 169–203.
32 See Becker, "Comparative Study," 108–09 for a thoughtful discussion of this concept.

its auditors to know?³³ How can we know the Jews and Christians it references? This is a particularly tricky question when we also consider the Qur'ān's rhetoric and style.³⁴ In other words, to the extent that the Qur'ān is polemical, or carrying a particular message – does it intend, or expect, a literal reading of its allusions to Jews/Christians? Further, as much exegesis was both chronologically and geographically removed from the qur'ānic milieu, the later exegetical glosses better reflect the milieus of the exegetes than the first auditors' reception of the qur'ānic revelations. If, however, we attempt to read the Qur'ān from a Late Antique (rather than later Islamic) perspective, we might be able to gain insight into the Judaism and Christianity the Qur'ān presumes its auditors to know, thereby illuminating our understanding of the qur'ānic message – while also expanding our perceptions of the involvement of "Arabs" in Late Antiquity.

The complexities of Late Antiquity are of particular interest for understanding the qur'ānic ambivalence towards monks and monasticism.³⁵ Although Islamic tradition has generally understood the Qur'ān as criticizing monasticism as an "innovation" (cf. Qur'ān 57:27), it also contains words of praise for Christians on account of monks and priests who are humble (Qur'ān 5:82). But, in Qur'ān 9:30–34, Christians are said to take their monks and *aḥbār* as "gods besides God," and these two groups are accused of taking the wealth of people. As these passages that allude to the misconduct of religious authorities and, arguably, their abuse

---

33 Contemporary scholarship on the Qur'ān, as well as classical exegetes, posit a range of possibilities. For an overview of the literature, see the aforementioned *EQ* articles on "Jews and Judaism" and "Christians and Christianity." Possible connections to "rabbinic" Judaism, Jewish-Christianity, Gnosticism, as well as the three well-known Christian divisions resulting from Chalcedon: Melkites, Jacobites and Nestorians (particularly in their Syriac manifestations), have all been posited. Other possibilities, such as a lingering late remnant of Jewish-Christianity, have also been posited.
34 See, e.g. the *EQ* articles "Rhetoric of the Qur'ān" and "Language and Style of the Qur'ān."
35 The multivalent qur'ānic estimations of monks and monasticism also resonates with Late Antique trends in Christian ecclesiastical circles. For, prior to the rise of Islam, the ecclesiastical hierarchy attempted to regulate monastic communities and individual ascetics—as at the Council of Chalcedon (451 CE) and that of Seleucia-Ctesiphon (481 CE). Not surprisingly, such regulations do not appear to have been met with universal approval by all Christian communities. As such, might the Qur'ān's first auditors (in contrast to later Islamic tradition) have heard echoes of Late Antique Christian debates over monasticism (rather than a blanket condemnation of the institution) in its allusions to the "innovation" of *rahbāniyya* and the mixed praise and criticism of *ruhbān* (Qur'ān 5:82–9:31/34). As Christian theological disputes, such as those that both led up to, and followed from, the Council of Chalcedon (451 CE) are found, for example, in qur'ānic references to Jesus and Mary, the Qur'ān might also reflect a knowledge in its milieu of the multiple faces of monasticism that both western and eastern Church councils of Late Antiquity also addressed.

of power sandwich the indeterminate reference to people who "wish to destroy the light of God with their mouths," might this passage also refer to a practice of religious authorities? For this charge parallels criticisms of (Hellenic) scholasticism practiced by Jewish and Christians scholars found in some Syriac Christian monastic literature that is similar to the tension between the Torah and *midrash* in Jewish tradition. For, as Becker has demonstrated, rather than extolling the Hellenic scholasticism popular in certain Jewish and Christian academic circles (such as that associated with the academies at Nisibis), this literature promotes "Egyptian"-type monasticism, which emphasizes silence and solitude. But, not all monastics were critical of the Hellenized schools; the qur'ānic accusation, then, against *aḥbār* and *ruhbān* found in Qur'ān 9:30–34 may reflect familiarity with the criticisms that solitaries in the desert had of their brethren who associated with the schools. Further, the term *aḥbār* has been glossed both as Christian and as Jewish scholars. Such a reading is in keeping with an account found in the *History of Mar Aba*, in which the holy man, prior to his conversion, met an east Syrian "school man" – but could not tell whether he was a Jew or a Christian![36]

Especially if the Qur'ān is understood as addressing people familiar with the solitude of the desert, might its charge of people wishing to "destroy the word of God with their mouths" reflect a sympathetic familiarity with an Evagrian criticism of the approaches to scripture found in both Jewish and Christian "schools" in the more settled urban areas – of Mesopotamia or, possibly, even in Arabia itself (e.g., in Najrān)? Although long studied as being far removed from the Hellenized world of Late Antiquity, the Qur'ān and the sixth-century Arabian peninsula are increasingly understood as knowledgeable of, and connected to, neighboring civilizations. Living far to the north of the peninsula, the sixth century Jacob of Serugh[37] was aware of the martyrdom of Christians in Najrān (southern Arabia)[38] in his own lifetime. And, a little over a century later, Seleucia was conquered by Arab/Muslim armies (637). Furthermore, the Qur'ān contains numerous allusions to events and figures familiar from Late Antique history and literature – not just from the Bible.[39] But, while the nature and extent of Christianity and Judaism in

---

36 Discussed in Becker, "Comparative Study," 108–9.
37 For more on Jacob of Serugh and the Qur'ān, see Sidney H. Griffith, "Christian Lore and the Arabic Qur'ān: The 'Companions of the Cave' in *Surat al-Kahf* and in Syriac Christian Tradition," in *The Qur'an in Its Historical Context*, ed. Gabriel S. Reynolds (London: Routledge, 2008), 109–37.
38 For a recent and thorough overview of eastern Christianity at the rise of Islam, see Lucas Van Rompay, "Society and Community in the Christian East," in *The Cambridge Companion to the Age of Justinian*, ed. Michael Maas (Cambridge: Cambridge University Press, 2005), 239–66.
39 See the excellent discussion in Griffith, "Christian Lore." Other contributions in Reynolds's volume also touch on this topic.

Late Antique Arabia has been the object of scholarly interest,[40] both from within and outside of Islamic tradition, there is as yet no scholarly consensus as to which Jews or Christians the Qurʾān expected its auditors to know. Reading the Qurʾān in the light of the events and literature of Late Antiquity might, however, help us to hear it as its first auditors may have. Such a reading may shed light not only on the involvement of the Qurʾān and its community in Late Antiquity, but on the Judaism and Christianity known to, if not in, Arabia.

The preceding has explored ways in which qurʾānic allusions to scriptural corruption, as well as its ambivalence towards monks and monasticism, resonate with Late Antique Jewish and Christian controversies. In particular, Qurʾān 9:32's allusion to the desire to "extinguish the light of God with their mouths" was examined in the light of Late Antique Jewish and Christian ambivalence towards (Hellenized) scholastic approaches to scripture, with particular attention to monastic preference for silent meditation on, and criticisms of scholastic investigation into, and disputes over, revealed texts.

References

Camilla Adang, *Muslim Writers on Judaism and the Hebrew Bible: From Ibn Rabban to Ibn Hazm* (Leiden: Brill, 1996).
Adam H. Becker, "The Comparative Study of 'Scholasticism' in Late Antique Mesopotamia: Rabbis and East Syrians," *Association of Jewish Studies Review* 34 (2010): 91–113.
Adam H. Becker, *Fear of God and the Beginning of Wisdom: The School of Nisibis and the Development of Scholastic Culture in Late Antique Mesopotamia* (Philadelphia: University of Pennsylvania Press, 2006).
Sheila S. Blair, "Writing and Writing Material," in *The Encyclopaedia of the Qurʾān*, ed. Jane D. McAuliffe (Leiden: Brill, 2005), 5:558–59.
Daniel Boyarin, *Border Lines: The Partition of Judaeo-Christianity* (Philadelphia: University of Pennsylvania Press, 2004).
François De Blois, "Naṣrānī (Ναζωραῖος) and Ḥanīf (ἐθνικός): Studies on the Religious Vocabulary of Christianity and of Islam," *Bulletin of the School of Oriental and African Studies, University of London* (2002): 1–30.
Ignazio di Matteo, "Il 'tahrif' od alterazione della Bibbia secondo i musulmani," *Bessarione* 38 (1922): 64–111, 223–60.
Emran I. El-Badawi, "From 'Clergy' to 'Celibacy': The Development of Rahbāniyyah between the Qurʾān, Ḥadith and Church Canon," *Al-Bayan: Journal of Qurʾān and Ḥadīth Studies* 11/1 (2013): 1–14.

---

[40] See, for example, Robert G. Hoyland, *Arabia and the Arabs: from the Bronze Age to the Coming of Islam* (London: Routledge, 2001). His more recent contribution to the *Oxford Handbook on Late Antiquity*, "Early Islam as a Late Antique Religion," provides a solid overview of this topic.

Emran I. El-Badawi, *The Qur'an and the Aramaic Gospel Traditions* (New York: Routledge, 2013).

Edmund Beck, ed., *Des heiligen Ephraem des Syrers Sermones de Fide*, Corpus Scriptorum Christianorum Orientalium 212 (Leuven: Peeters, 1961).

Reuven Firestone, "The Failure of a Jewish Program of Public Satire in the Squares of Medina," *Judaism* 46/4 (1997): 439–52.

Sidney H. Griffith, "Christian Lore and the Arabic Qur'an: The 'Companions of the Cave' in *Surat al-Kahf* and in Syriac Christian Tradition," in *The Qur'an in its Historical Context*, ed. Gabriel S. Reynolds (London: Routledge, 2008), 109–37.

Sidney H. Griffith, "Christians and Christianity," in *The Encyclopaedia of the Qur'ān*, ed. Jane D. McAuliffe (Leiden: Brill, 2001), 1:307–16.

Sidney H. Griffith, "Images of Ephraem. The Syrian Holy Man and his Church," *Traditio. Studies in Ancient and Medieval History, Thought and Religion* 45 (1989–90): 7–33.

Sidney H. Griffith, "Al-Nasara in the Qur'an: A Hermeneutical Reflection," in *New Perspectives on the Qur'an: The Qur'an in Its Historical Context 2*, ed. Gabriel S. Reynolds (New York: Routledge, 2012), 1–38.

Sidney H. Griffith, "Syriacisms in the Arabic Qur'ān: Who Were 'Those who Said 'Allāh is Third of Three' According to al-*Mā'idah* 73?" in *A Word Fitly Spoken: Studies in Mediaeval Exegesis of the Hebrew Bible and the Qur'ān; Presented to Haggai Ben-Shammai*, ed. M. Bar-Asher et al. (Jerusalem: Hebrew University Press, 2007), 83–110.

Robert G. Hoyland, *Arabia and the Arabs: from the Bronze Age to the Coming of Islam* (London: Routledge, 2001).

Robert G. Hoyland, "Early Islam as a Late Antique Religion," in *The Oxford Handbook of Late Antiquity*, ed. Scott Fitzgerald Johnson (Oxford: Oxford University Press, 2012), 1054–77.

'Imād al-Dīn Ismā'īl b. 'Umar Ibn Kathīr. *Tafsīr al-Qur'ān al-'aẓīm*, ed. 'Abd al-'Azīz Ghunaym et al. (Cairo: n.p., 1971).

Aryeh Kofsky "Eusebius of Caesarea and the Christian-Jewish Polemic" in *Contra Iudaeos: Ancient Polemics between Christians and Jews*, ed. Ora Limor and Guy Stroumsa, Texts and Studies in Medieval and Early Modern Judaism 10 (Tubingen: Mohr, 1996), 59–84.

Louise Marlow, "Scholar," in *The Encyclopaedia of the Qur'ān.*, ed. Jane D. McAuliffe (Leiden: Brill, 2004), 4:537–40.

Abū l-Ḥasan Muqātil b. Sulaymān al-Balkhī, *al-Tafsīr*, ed. 'Abdallāh Maḥmūd Shiḥāta (Cairo: n.p., 1980–87).

Angelika Neuwirth, "Qur'an and History – A Disputed Relationship: Some Reflections on Qur'anic History and History in the Qur'an," *Journal of Qur'anic Studies* 5/1 (2003): 1–18.

Angelika Neuwirth, "Rhetoric of the Qur'ān," in *The Encyclopaedia of the Qur'ān*, ed. Jane D. McAuliffe (Leiden: Brill, 2004), 4:461–76.

Arye Olman, "'The Jews Distorted Torah': An Attempt to the Moslem Claim," in *Proceedings of 15th Conference of Jewish Studies in FSU 'Sefer'* 2 (2008): 90–100.

Fakhr al-Dīn al-Rāzī, *al-Tafsīr al-kabīr (Mafātīḥ al-ghayb)*, ed. Muḥammad Muḥyī l-Dīn 'Abd al-Ḥamīd (Cairo: n.p., 1933).

Gabriel S. Reynolds, "On the Qur'anic Accusation of Scriptural Falsification (taḥrīf) and Christian Anti-Jewish Polemic," *Journal of the American Oriental Society* 130 (2010): 189–202.

Lucas Van Rompay, "Society and Community in the Christian East," in *The Cambridge Companion to the Age of Justinian*, ed. Michael Mass (New York: Cambridge University Press, 2005), 239–66.

Uri Rubin, "Children of Israel," in *The Encyclopaedia of the Qur'ān*, ed. Jane D. McAuliffe (Leiden: Brill, 2001), 1:303–7.

Uri Rubin, "Jews and Judaism," in *The Encyclopaedia of the Qur'ān*, ed. Jane D. McAuliffe (Leiden: Brill, 2003), 3:21–34.

Moshe Sharon, "People of the Book," in *The Encyclopaedia of the Qur'ān*, ed. Jane D. McAuliffe (Leiden: Brill, 2004), 4:36–43.

Abū Ja'far Muḥammad b. Jarīr al-Ṭabarī, *Jāmi' al-bayān 'an ta'wīl āy al-Qur'ān*, ed. Aḥmad Sā'īd 'Alī et al. (Cairo: n.p., 1954–57).

Paul E. Walker, "Knowledge and Learning," in *The Encyclopaedia of the Qur'ān*, ed. Jane D. McAuliffe (Leiden: Brill, 2003), 3:100–104.

Holger Michael Zellentin, *The Qur'ān's Legal Culture: The Didascalia Apostolorum as a Point of Departure* (Tübingen: Mohr Siebeck, 2013).

Paul E. Walker
9 Techniques for Guarding and Restricting Esoteric Knowledge in the Ismaili *Daʿwa* during the Fatimid Period

9.1 Introduction

In Islamic doctrine the classical point of departure for discussions of what knowledge can or cannot be known, and can and cannot be revealed openly and thus must be carefully controlled, is Qur'ān 3:7, which states fairly explicitly that the sacred book contains verses some of which are clear and unambiguous but others that are ambiguous. The latter require some form of interpretation (*taʾwīl*).[1] The problem in front of us is how to determine who knows what the interpretation is and who might be authorized to speak about it. A common understanding of the same passage would claim that only God knows and humans do not and cannot. However, the teaching of the Philosophers in Islam, most particularly Ibn Rushd (Averroes), would insist that the elite scholars in the community, those capable of demonstrative thought, do have the requisite knowledge and should make such interpretations. They are the ones in the category of "those firmly grounded in knowledge" mentioned in this very qurʾānic text. They must not, however, speak about them to the general public but rather restrict their communications solely to those like themselves. Common folk, who are not capable of theoretical knowledge, are in fact harmed by being exposed to such abstractions. Here is what Ibn Rushd says:

> When something of these allegorical interpretations is expressed to anyone unfit to receive them – especially demonstrative interpretations because of their remoteness from common knowledge – both he who expresses it and he to whom it is expressed are led into unbelief. The reason for that in the case of the latter is that allegorical interpretation comprises two things, rejection of the apparent meaning and affirmation of the allegorical one; so that the

---

[1] The whole text of this verse runs as follows: "It is He who sent down to you the book, wherein are verses clear that are the essence of the Book, and others that are ambiguous. As for those in whose hearts is swerving, they follow the ambiguous part, desiring dissension, and desiring its interpretation (*taʾwīl*). But no one knows its interpretation (*taʾwīl*) save only God and those firmly grounded in knowledge; they say, 'We believe in it.' All is from our Lord; yet no one remembers but men possessed of minds." Needless to say, there are many ways to translate this verse and punctuation can change and with it the meaning. It is also much discussed in the tradition.

https://doi.org/10.1515/9783110596601-010

apparent meaning is rejected in the mind of someone who can only grasp apparent meanings, without the allegorical meaning being affirmed in his mind, the result is unbelief ....[2]

His order not to speak about such knowledge to the general public is quite explicit:

> One must not speak about those things concerning which the Holy Law is silent; the masses must learn that human understanding is not sufficient to treat these problems, and must not go beyond what the teaching of the Holy Law explains in its texts, since this is teaching in which all can participate and which suffices for the attainment of their happiness. And just as the physician investigates the measure of health which agrees most with the healthy for the preservation of their health, and with the sick for the curing of their illness, so the Lord of the Holy Law instructs the masses only in so far as is needed for their acquisition of happiness.

Among the Ismaili Shi'a, and some other branches of the Shi'a as well, it is standard to claim that knowledge of the inner, esoteric meaning of scripture, as in the verses of the Qur'ān that are admittedly ambiguous, fall into the domain of the imams. The imams, commencing with 'Alī b. Abī Ṭālib, as the chosen successor to the Prophet Muhammad, and then running through his progeny one by one, know the true meaning of all scripture and the Law based on it. It is they who are "those firmly grounded in knowledge," again with reference to the qur'ānic text. What is obvious in it is its literal meaning, its exoteric dimension (ẓāhir), whereas the truth (or truths) behind the outward sense, its esoteric side (bāṭin), is concealed and is thus not available to ordinary Muslims. They do not and cannot know the interpretation solely as the result of personal effort. The truth comes only from the Imam.

Here then is an important distinction, one that needs to be cited here at the beginning because it involves an essential difference that sets Ismaili doctrine apart and in clear contrast to the Philosopher's limited esotericism. The latter is what Ibn Rushd calls the "knowledge of the steadfast."[3] He and the Philosophers like him, would argue that the allegorical interpretation of scripture, which becomes essential when its apparent meaning appears to conflict with reason, can be discovered rationally by the extension of the significance of an expression from real to metaphorical but without forsaking standards of metaphor in

---

2 Ibn Rushd, *Faṣl al-maqāl*, trans. George F. Hourani, "The Decisive Treatise," in *Averroes On the Harmony of Religion and Philosophy* (London: Luzac, 1976), 66.
3 "But all this is the knowledge of those who are steadfast in their knowledge, and this must not be written down and it must not be made an obligation of faith, and therefore it is not taught by the Divine Law. And one who mentions this truth where it should not be mentioned sins, and one who withholds it from those to whom it should be told sins too."

the language.⁴ For the Ismailis there is no such restriction; esoteric knowledge (the *bāṭin*) need not be so confined. For them, as explained by their principal authorities – a selection of writers from the Fatimid period whose works we now possess – the matter is much more complicated. While the operation of interpretation as employed by the Philosophers (and others) is quite possible even for these Ismailis, there are for them more layers of explanation than that. All items of doctrine mentioned in scripture and the Law can have symbolic value. In fact all aspects and parts of the natural realm, the universe at large and in particulars, exist in parallel with corresponding religious values, ranks and meanings. Each may have a symbolic counterpart in the realm of faith. Each symbolizes something else, something of special religious meaning, often of particular significance in Ismaili propaganda and its appeal, in Arabic the *daʿwa*. There may be no logical, that is, rational, connection between the symbol and what is symbolized by it. We have here knowledge that is truly esoteric. It is not merely difficult to understand, but is genuinely beyond the capacity of humans of whatever class because it belongs solely to the imams and those authorized to represent them.

The Shiʿa believe that failure to acknowledge the living imam, the imam of the time, leads to perdition and hellfire. That condition results in part from not having access, as a consequence, to the knowledge required for entry into eternal bliss and salvation. One source⁵ explains that the human soul commences its existence as the first perfection of the body to which it is attached. If it has access to a teacher who can provide it with the truth, i.e. the true meaning of religion, it moves toward an end in which it will have acquired a second perfection and thus achieved perpetual felicity in the abode of the everlasting. The teacher is the imam. Lacking access and failing to adhere to the cause of the imam and his line implies doom and damnation. Thus knowledge of the esoteric dimension of the truth is key and that can only come from the *daʿwa* and its agents, the *dāʿīs*, who control access to it.

Of the several techniques used to restrict access, two are especially important. One is the oath of covenant administered to each and every adept who wishes to

---

4 "The meaning of *taʾwīl* (allegorical interpretation) is the extension of the significance of an expression from real to metaphorical significance without forsaking therein the standard metaphorical practices of Arabic, such as calling a thing by the name of something resembling it or a cause or consequence or accompaniment of it, or other things such as are enumerated in accounts of the kinds of metaphorical speech" (Ibn Rushd, "The Decisive Treatise," trans. Hourani, 50).

5 This form of the doctrine is that of the famous early eleventh century Ismaili *dāʿī* Ḥamīd al-Dīn al-Kirmānī on whom in general and this concept of the soul in particular, see Paul Walker, *Ḥamīd al-Dīn al-Kirmānī: Ismaili Thought in the Age of al-Ḥākim* (London: I. B. Tauris, 1999).

join the cause. We possess in fact copies of an exact text employed and descriptions of a kind of ceremony in which the neophyte swears his or her allegiance. One provision in it makes sure that all knowledge that is imparted to the novice is never revealed to others without explicit permission from senior officials in the *da'wa*. No such material is ever to be shared with those to whom it does not belong (i.e., are not members of the *da'wa* in good standing). The other method of control involves payments of various tithes and fees of one kind or another, the payment of which is often a test of the new member's sincerity and firm conviction. But it is only after such payments that the agent of the *da'wa* is allowed to divulge to the new initiate the esoteric knowledge he or she possesses. Exactly how this process works and under what circumstances is especially evident in a newly uncovered treatise from the earliest Fatimid period explaining the ways such payments can lead to esoteric knowledge, which, in turn, supplies the keys to salvation.

Thus in the rest of this paper it is essential to explore and explain in reasonable detail these two aspects (the oath and financial payments) of the Ismaili methods of controlling access to the interpretive (*ta'wīl* based) knowledge by the imam and the agents acting on his behalf, all the while stressing that according to the doctrine behind this policy, its purpose is the ultimate salvation of those who subscribe to this system.

9.2 The Oath of Allegiance

The Ismaili oath of allegiance is now fairly well documented. We have examples, whole or partial, from various periods and, since it is still used in one form or another by modern Ismailis, we can be reasonably certain that it is authentic. All persons who joined the *da'wa* swore such an oath and to the covenant it entailed. It is unclear, however, whether or not all Ismailis took the oath, which would imply that they had become members of the *da'wa*. Were all regarded as members of the *da'wa*, meaning any and all who had responded to the appeal? That is less than clear. But we do understand that the basic level in the *da'wa* was that of the novice, the *mustajīb*, "the one seeking answers." To judge from the variety of *majālis*, the teaching and exhortation sessions convened by the agents, of which we possess several accounts and examples, they included women as well as men.[6]

---

[6] We have accounts of these sessions specifically devoted to the instructing of women in Ismaili teachings. For one such example see Wilferd Madelung and Paul E. Walker, *The Advent of the Fatimids: A Contemporary Shi'i Witness* (London: I. B. Tauris, 2000), 168–70 (English). On this

The ceremony itself is described as solemn and certainly not lightly undertaken.[7] The *dāʿī* administering the oath must ascertain carefully and surely the sound and sincere purpose of the new adept. There are ritual steps: citation of qurʾānic precedence, extirpation of previous beliefs, fasting, and prayer. Al-Naysābūrī, a leading *dāʿī* in the time of al-Ḥākim (r. 996–1021), explains the process as follows:

> when he [the novice] has been broken down and wants to take the oath, the rule is to take it of him after he has fasted three days. Both the *dāʿī* and the novice should be in a pure state. Each should pray two prostrations so as to be at the most pure. Next he begins with thanking God the highest and praising Him, asking for His blessing on His messengers and on the pure imams. He takes of him the oath of covenant to God, His angels, His messengers, the oath to the legatees and the pure imams, and an oath to the imam of his own time, may the blessing of God be on them. He pledges to him allegiance, as is the rule for that in the book of covenant. He swears that he believes in God, and in His angels, His messengers, the pure imams from the legatee to the imam of the time, may the blessings of God be on them all, and that he will uphold the external and the internal and that he will support the imam of his time and will not forsake him, that he will not reveal any secret of the faith to a person not worthy of it or to anyone who has not sworn the oath of covenant, that he will not betray any of the brethren of believers who have joined him in swearing the oath, that he will treat as a friend those who have accepted the imams and as an enemy those who are enemies of theirs, that he will stay away from their enemies, that he will offer good counsel on behalf of God and His representative, upon whom be peace, and that, if he should go back on his oath, there will apply to him what applies to those who rescind or violate an oath, that he will appeal for the imam of their time, ascribe knowledge to him and not a letter of that to himself.[8]

Note in particular the requirement of secrecy. A person given answers, i.e., the novice, is not to divulge those answers to anyone "not worthy of them."

Here is another version of the oath[9]:

---

institution in general see Walker, "Fatimid Institutions of Learning," *Journal of the American Research Center in Egypt* 34 (1997): 179–200, at 182–86; reprinted in *Fatimid History and Ismaili Doctrine* (Aldershot: Ashgate, 2008), 7–15.

7 An important addition to the examples that follow is found in the translation of the memoire of Ibn al-Haytham by Madelung and Walker, *The Advent of the Fatimids*, 95. There the author describes being "summoned to the faith" and taking the "oath." The date of this event was Spring of 909 and it is therefore one of our earliest accounts.

8 Paul Walker and Verena Klemm, ed. and trans., *A Code of Conduct: The* Mūjaza al-kāfiya *of al-Naysābūrī, Critical Edition of the Arabic and Complete English Translation with Notes and Introduction* (London: I. B. Tauris, 2011), 61.

9 Al-Maqrīzī (*Khiṭaṭ* 2:318–20) gives us a version of the Ismaili oath that he has taken from the work of Akhū Muḥsin (on which see Heinz Halm, "The Ismaili Oath of Allegiance (*ʿahd*) and the 'Sessions of Wisdom" [*majālis al-ḥikma*] in Fatimid Times," in *Mediaeval Ismaʿili History and Thought*, ed. Farhad Daftary (Cambridge: Cambridge University Press, 1996), 75–115, at 94–96). Al-Maqrīzī calls this section of his work "A description of the oath that is taken from the one summoned." The following translation is that of Halm as it appears in that article (in English translation).

The *dāʿī* shall say to the person to whom he administers the oath (*ʿahd*): You impose on yourself God's oath (*ʿahd*), compact (*mīthāq*) and obligation (*dhimma*), as well as the obligation of God's envoy, of his prophets, angels, books and envoys, the same pledge, contract and obligation which He entered into with His [earlier] prophets: that you will keep secret what you will hear and what you have already heard, what you know and what you will learn, what knowledge you have and what knowledge you will yet acquire concerning myself and the one who dwells in this city as representative of the Lord of truth, the imam for whom – as you know – I declare myself, and to whose committed adherents I openly and honestly belong .... Thus you must reveal nothing of it, neither little or much, nor in allusions, except those things about which I myself or the person responsible dwelling in this city explicitly allow you to speak, so that in this matter you must only act according to our command, which you must not contravene and to which you must add nothing. ...

You must be friendly to the friends of God and hostile to His enemies and observe God's duties and customs (*sunan*) as well as the customs of His prophet, in the exoteric and in the esoteric sense (*ẓāhiran wa bāṭinan*), both openly and in secret. All this is confirmed by this oath, rather than being invalidated by it; it is corroborated rather than cancelled by it, brought closer rather than removed, strengthened rather than weakened, imposed as a duty rather than repealed, clarified rather than obscured. This is equally true of the exoteric and the esoteric, and of all that the prophets have revealed of the Lord, according to the manifest conditions contained in this oath. – If you pledge to observe all this, then say: 'yes.' Thereupon the summoned one says, 'yes.'

Then the *dāʿī* says to him: The security and warrant for it are that you will not reveal any of the things which you have pledged by this oath, neither in our lifetime nor after our death, neither in anger nor in a contented mood, neither from desire nor from fear, neither in distress nor when at ease, neither out of greed nor out of need – may God assume the protection and guarantee for it! – according to the manifest conditions contained in this oath! .... So never betray either God or His friend, either us or one of our brothers and friend or anyone of whom you know that he belongs to us, whether for family reasons or for money, or else because of an opinion, an oath or an agreement which you might interpret as invalidating [this oath].

If you do anything of the kind, although you know that in so doing you violate [the oath], which you remember exactly, then you renounce God the Creator of heaven and earth, ... You renounce His earlier and later envoys, His angels who are close to Him, ... You renounce the Torah, the Gospel, The Psalter and the Wise Admonition (3:58 and elsewhere), every religion sanctioned by God in the approaches of the world to come, and every servant who is pleasing to God. You leave the party of God and his saints, to God's unconcealed disappointment, but He will soon bring retribution and punishment on you, and you will walk into the fire of *jahannam* [Hell], in which there is no mercy. You renounce the power and strength of God and rely on your own power and strength – so may there be on you the same curse of God with which He cursed the Iblīs, which barred paradise to him and assigned him to the fire forever!

If you violate any part of all this, you will some day appear before an angry God. ... And if you violate it, then everything you have acquired during the time of your violation shall be given as alms (*ṣadaqa*) to poor and miserable people who are not related to you by blood. This will bring you no reward from God, ... If you violate any part of it, all your slaves, male or female, who are in your possession and whom you will take into your service until the hour of your death will be free in the sight of God; all your wives, including those whom you will marry up to the hour of your death, will be divorced ....[10]

From these examples it is relatively easy to see that the oath serves two main purposes. One is to ensure the absolute loyalty of each new member or adherent and the other, more germane for us, is to guard and protect and thus control access to the esoteric knowledge imparted in the course of the da'wa's appeal and its instruction. Note in particular the following language of the first oath. The new adept swears that "he will support the imam of his time and will not forsake him, that he will not reveal any secret of the faith to a person not worthy of it or to anyone who has not sworn the oath of covenant." Or that of the second version: "Thus you must reveal nothing of it, neither little or much, nor in allusions, except those things about which I myself or the person responsible dwelling in this city explicitly allow you to speak, so that in this matter you must only act according to our command, which you must not contravene and to which you must add nothing." ... "If you do anything of the kind, although you know that in so doing you violate [the oath], which you remember exactly, then you renounce God the Creator of heaven and earth.... You leave the party of God and his saints, to God's unconcealed disappointment, but He will soon bring retribution and punishment on you, and you will walk into the fire of *jahannam* [Hell], in which there is no mercy. You renounce the power and strength of God and rely on your own power and strength – so may there be on you the same curse of God with which He cursed the Iblīs, which barred paradise to him and assigned him to the fire forever!"

9.3 Payments of Dues and Alms

Payments made to gain access to esoteric knowledge – what is sometimes called "the interpretive sciences" (*al-'ulūm al-ta'wīliyya*) – and the status accorded by membership at a higher rank are fairly clear in general terms. There were set fees for attendance at the weekly lessons (*majālis*); and additional dues for other occasions. Some, such as the regular payment called the "fee for confidential discourse" (*najwa*) is qur'ānic in origin, although not a standard institutionalized Islamic practice except for Shi'a, especially the Ismailis and a few others.

---

[10] Halm, "The Ismaili Oath." This particular version of the oath as trans. by Halm, appears on pp. 95–97. The rest of this article is also highly important for our subject.

However, the exact correlation between payment and access to esoteric knowledge, that is, the interpretive sciences, might have appeared vague and imprecise from the evidence previously available. But new material, most particularly a critically important text from the very beginnings of the Fatimid caliphate, reveals in fairly explicit language how the connection between the two operated within the *da'wa*. The Ismaili movement certainly existed and had spread dramatically prior to the founding of the state in 909 in North Africa but it, until then, remained largely underground and out of view. Our knowledge about it is therefore limited. Even afterward, when the Ismailis actually governed substantial territory, their doctrine, teaching and practice tended to be kept tightly restricted, in part to prevent hostile attacks from Sunni and other anti-Shi'i Muslims who had now either become subjects of the new state or enemies outside. Moreover, the literature produced over the two and half centuries of the Fatimid caliphate remained throughout under watchful guard and out of the hands of detractors. That situation which has made the investigation of our topic often quite difficult is slowly changing with the recovery more recently of long concealed texts. And although much still needs to be done, some vitally important new material has emerged and more is on the way.

For our purposes the most important item is a work, a lengthy letter, composed by the brother of the famous Fatimid agent in North Africa who was responsible for the creation of the state, Abū 'Abdallāh al-Shī'ī.[11] The brother, Abu'l-'Abbās, wrote this document to explain carefully the fiscal duties of a loyal member of the Ismaili system as an adherent of the imam. Near the beginning he makes the connection between payments and knowledge clear:

> With the payment [one makes] without having being forced or compelled, rendering that offering out of the goodness of one's own self, God will cleanse thereby his spirit and purify his money, and thus it will be lawful for his agent [*dā'ī*] and mentor to reveal to him the interpretive (*ta'wīlī*) sciences and make known to him the truths hidden from the enemies of God's religion.

Such payments are in part set up as an aspect of the duty of giving alms. But, in the Ismaili context, there is more to it. As our text says:

> The believer pays what he pays on the measure of his sincerity and in accord with what his *dā'ī* determines for him in order to test him. If he pays that once, he has fulfilled the basic requirement of the religion and thus fulfills the necessary obligation, by his fulfillment of

---

11 Wilferd Madelung and Paul E. Walker, eds., *Two Works from the Earliest Fatimid Da'wa in North Africa: Sermons by Abū 'Abdallāh al-Shī'ī and a Letter by His Brother Abu'l-'Abbās on Fiscal Obligations to the Imam*
(forthcoming).

which he distinguishes himself from the people of outward meaning [Sunni Muslims and other non-Ismailis] and he departs from their ranks. If he pays a second time, his *dāʿī* knows the goodness of his intention and the firmness of his certainty and then he reveals to him the *secrets of the interpretation*. If he pays out a third time, his status with God's guardian rises and his rank similarly rises among the believers.

When the *dāʿī* puts the novice to the test and the novice bears up patiently through the ordeal, he is allowed to initiate him in the keys to knowledge. If he flees and does not bear patiently the ordeal, initiation in knowledge is forbidden.

If he gives what he is ordered to give by his *dāʿī* and guards against that in which there are the people of confusion and discord who do not know the interpretation of the books of God. If he does that, the way is made easy, his knowledge increases and his prestige rises with the believers and with God's guardian. His rank increases and his station rises. "But the one who is miserly, keeps for his own self, and considers the good to be false" (Qurʾān 92:8–9), that is, he who is miserly in respect to the obligation he has been commanded to observe and does not produce it, regarding himself as above having to pursue that in which there is his very life, and who declares false the interpretation God set forth to expand the understanding of the believers. "So for him We will ease the way to adversity" (Qurʾān 92:10) and poverty, thus cutting off for him the substance of the knowledge by means of which is his salvation.

God orders him [the *dāʿī*] to exam the believers who seek the benefits of the religion to test out their secrets. If they bear the trial with patience, it is licit for the teacher to initiate them and raise them in the interpretive sciences.

At this point the critical importance of the term *taʾwīl* should have become obvious. It means various things: allegorical interpretation, or other forms of interpretation, of qurʾānic verses or legal texts, this being only one facet or aspect of a complex network of methods of understanding it and its significance. This topic itself is vast in Islamic discourse. For the Ismailis it is likewise a matter of much debate and yet is clearly essential to its doctrinal program.[12] What is outlined above only begins to probe the range of issues involved. At one end of a broad spectrum are the symbolic interpretations of religious texts, either separate particular passages or even whole sections. At the other end are what we moderns would call the natural sciences. Interpretation here deals with the uncovering or discovering of the reality behind, say, a physical object. What is it made of, its constituent elements? What can be made of it, as in wood that becomes a table or a chair?

---

[12] The most recent major study of *taʾwīl* in the Ismaili context is David Hollenberg, *Beyond the Qurʾān: Early Ismāʿīlī Taʾwīl and the Secrets of the Prophets* (Columbia: University of South Carolina Press, 2016).

Ismaili writings, that is, the literature produced by the agents of the *da'wa*, from the Fatimid period 909 to 1171, are substantial, perhaps two hundred works in all. Many are small, short treatises of one kind or another, but there are also major items of considerable length. A century ago we would have known little about what has survived and what might be recovered. In part our ignorance was a deliberate result of Ismaili secrecy, of its stated policy of guarding its doctrine from outside scrutiny. The *da'wa* controlled access to these works and only rarely did non-Ismailis see them. But now most of those we know about are becoming accessible, either in manuscript or print, with a few even translated into a language other than Arabic or Persian. That then brings up the question of how much of this substantial corpus of *da'wa* literature is to be classified as *ta'wīl* interpretation. Given the wide reach of subjects discussed in these texts, where do we find in it *ta'wīl*? And what portion of it falls under the rule of secrecy. Is all of it protected by the oath?

For modern scholars gaining access has often been difficult. Those who preserve these texts regard them as, for the most part, restricted. Outsiders should still not be allowed to read them. Even members of these communities need special permissions only granted after extensive training and education. But, as it has happened, copies have found their way piecemeal into our hands, few at first but many more later. Thus we have gained access without agreeing to the oath or making the necessary payments of fees and dues. Most importantly we are not loyal adherents of the cause or properly trained in its doctrine. We have penetrated the wall of secrecy and uncovered what has for hundreds of years been hidden and carefully guarded.

A central issue here involves determining what the Ismaili tradition calls the *haqā'iq*, the "truths" or perhaps the "ultimate realities." These lie at the core of its beliefs; they are the most sensitive items of doctrine, that portion the uncontrolled revelation of which could cause substantial harm to the community. But for scholars outside it is extremely difficult to find the exact line between what can be shared and what must be hidden, since we have no one to guide us to it.

The corpus of *da'wa* literature certainly contains a significant number of works composed for a public purpose. They are therefore part of an ongoing debate by Ismailis with their various adversaries, either to refute an opponent or to attract new adherents. Such treatises are clearly not part of the *haqā'iq* and, rather than being restricted or guarded, are meant to engage the broader non-Ismaili community of scholars. A couple of examples here should suffice. They might include the legal materials produced by Qāḍī al-Nu'mān,[13] as well as his

---

13 Al-Qāḍī al-Nu'mān, *Da'ā'im al-Islām*, ed. Asaf A. A. Fyzee, Eng. trans. Ismail Poonawala, (Oxford: Oxford University Press, 2004).

numerous treatises of refutation of opponents,[14] or al-Kirmānī's *Lights to Illuminate the Proof of the Imamate*,[15] which was written to convince the Buyid wazir of Baghdad to convert to the Fatimid side.

But, about those that we are fairly sure were not to be made public, there still remain a question of the role in them of *ta'wīl*. Does a major work on the proofs of prophecy[16] (by al-Sijistānī) fall within the realm of *ḥaqā'iq* and thus require access solely via the system of oath, payments and training? From what we know at this point it appears that it does. Even if portions of it could be shared with outsiders, too much of it cannot and it thus remains, for the tradition, a secret text.

References

Heinz Halm, "The Ismaili Oath of Allegiance ('*ahd*) and the 'Sessions of Wisdom" (*majālis al-ḥikma*) in Fatimid Times," in *Mediaeval Isma'ili History & Thought*, ed. Farhad Daftary (Cambridge: Cambridge University Press, 1996), 75–115.

David Hollenberg, *Beyond the Qur'an: Early Ismā'īlī Ta'wīl and the Secrets of the Prophets* (Columbia: University of South Carolina Press, 2016).

Ibn Rushd, *Faṣl al-maqāl*, Eng. trans. G. Hourani, "The Decisive Treatise" in *Averroes On the Harmony of Religion and Philosophy*, ed. G. Hourani (London: Printed for the Trustees of the "E.J.W. Gibb Memorial" and pub. by Luzac & Co., 1976).

Wilferd Madelung and Paul Walker, eds. and trans., *The Advent of the Fatimids: A Contemporary Shi'i Witness* (London: I. B. Tauris, 2000).

Wilferd Madelung and Paul Walker, eds. and trans., *Two Works from the Earliest Fatimid Da'wa in North Africa: Sermons by Abū 'Abdallāh al-Shī'ī and a Letter by His Brother Abu'l-'Abbās on Fiscal Obligations to the Imam* (forthcoming).

Qāḍī al-Nu'mān, *Da'ā'im al-islām*, ed. A. A. Fyzee (Cairo: n.p. 1951–61); Eng. trans. Ismail Poonawala (Oxford: Oxford University Press, 2004).

Abū Ya'qūb al-Sijistānī, *Ithbāt al-nubuwwāt*, critical ed. Wilferd Madelung and Paul E. Walker, Intellectual Heritage of Islamic Civilization, Series I (Tehran: Ketāb-e Rāyzan, 2016).

Paul Walker, "To What Degree was Classical Ismaili Esotericism based on Reason as Opposed to Authority?" in *Esotérisme shi'ite, ses racines et ses prolongements*, ed. M. A. Amir-Moezzi, M. De Cillis, D. De Smet, O. Mir-Kasimov (Turnhout: Brepols, 2016), 493–505.

Paul Walker, *Early Philosophical Shiism: The Ismaili Neoplatonism of Abū Ya'qūb al-Sijistānī* (Cambridge: Cambridge University Press, 1993).

---

14 Al-Qāḍī al-Nu'mān, *Ikhtilāf uṣūl al-madhāhib*, trans. Devin Stewart in his *The Disagreement of the Jurists: A Manuel of Islamic Legal Theory* (New York: New York University Press, 2015).

15 *Master of the Age: An Islamic Treatise on the Necessity of the Imamate*, with a critical edition of the Arabic text of Ḥamīd al-Dīn al-Kirmānī's *al-Maṣābīḥ fī ithbāt al-imāma* (*Lights to Illuminate the Proof of the Imamate*), Full Translation, Introduction and Notes (London: I. B. Tauris, 2007).

16 Abū Ya'qūb al-Sijistānī, *Ithbāt al-nubuwwāt*, critical ed. Wilferd Madelung and Paul E. Walker, Intellectual Heritage of Islamic Civilization, Series I (Tehran: Ketāb-e Rāyzan, 2016).

Paul Walker, *Ḥamīd al-Dīn al-Kirmānī: Ismaili Thought in the Age of al-Ḥākim*. (London: I. B. Tauris, 1999).

Paul Walker, *Master of the Age: An Islamic Treatise on the Necessity of the Imamate* [includes a critical edition of the Arabic text of Ḥamīd al-Dīn al-Kirmānī's *al-Maṣābīḥ fī ithbāt al-imāma* (*Lights to Illuminate the Proof of the Imamate*), full translation, introduction and notes] (London: I. B. Tauris, 2007).

Author Index

Abegg, Martin G. 49–50
Abusch, Tzvi 22, 25
Adams, James N. 52, 57
Adang, Camilla 172
Aharoni, Yohanan 60
Alexander, Philip S. 59, 63, 98
Algra, Kempe 136
Ando, Clifford 87
Anipa, Kormi 47
Aptowitzer, Victor 98
Armstrong, A. Hilary 130, 138
Athanassiadi, Polymnia 130–131

Babut, Daniel 136
Bacher, Wilhelm 98
El-Badawi, Emran I. 174
Bagnall, Roger S. 52
Baker, Heather D. 23
Baltes, Matthias 128
Barjamovic, Gojko 13
Baumgarten, Albert I. 56
Bayley, Robert 48
Beaulieu, Paul-Alain 28
Beck, Edmund 179
Becker, Adam 177–180, 182
Bellman, Beryl L. 121
Ben Ezra, Daniel Stökl 61
Ben-Dov, Jonathan 48, 61, 63
Bernstein, Moshe 77
Berthelot, Katell 87, 98–103
Bezold, Carl 22
Bichler, Reinhold 29
Biscardi, Arnaldo 112
Black, Jeremy A. 18, 21, 25
Blain, Sheila 171
Blois, François De 174
Bok, Sissela 121
Bonazzi, Mauro 129
Borger, Rykle 24
Bourdieu, Pierre 65
Boyancé, Pierre 136
Boyarin, Daniel 176
Bremmer, Jan N. 122, 129, 135
Brenk, Frederick E. 131

Brickhouse, Thomas C. 156
Broek, Roelof van den 121, 126, 128
Bühler, P. 131
Burkert, Walter 136
Burnyeat, Myles 166
Buylaere, Greta Van 8

Cameron, Richard 48
Cancik, Hubert 137
Carter, Warren 103
Casanova, Angelo 110
Catan, John R. 137
Cioată, Maria 66
Clackson, James 54
Clancier, Philippe 29, 32
Claußen, Carsten 59
Cole, Stephen W. 27
Collins, John J. 50
Conde-Silvestre, Juan Camilo 47
Corrigan, Kevin 127
Cotton, Hannah M. 52, 58–59
Crawford, Harriet 29
Crawford, Sidnie White 60
Crespo, Manuel 128

Daftary, Farhad 190
Dahlvik, Julia 112
Danby, Herbert 91
David, Joseph 98
Davies, Richard 137
Davies, William D. 109, 114
DeConick, April D. 121, 122
Delcor, Matthias 53
Dias, Paula Barata 107
Dietrich, Manfred 8
Dijk, Jan J. A. van 32
Dillery, John 33
Dillon, John J. 124
Dohrmann, Nathalie B. 103
Dölger, Franz J. 126
Donaldson, James 124, 125
Donaldson, Terence L. 73
Dörrie, Heinrich 136

Edzard, Dietz O. 22
Eichler, Barry L. 26
Eliot, Simon 26
Elwolde, John F. 47
Endres, John C. 74
Erskine, Andrew 107
Eshel, Esther 59
Exum, J. Cheryl 72

Faivre, Xavier 32
Fales, F. Mario 21
Fales, Mario 25, 27
Feldman, Ariel 66
Feldman, Louis H. 108, 109, 115
Ferreira, José Ribeiro 107
Ferroni, Lorenzo 127
Fialho, Maria C. 106
Fialho, Maria do Céu 108
Fidanzio, Marcello 62
Fields, Weston W. 59
Fincke, Jeanette C. 27
Finkel, Irving L. 16, 24
Finkelstein, Louis 14–15, 88, 109, 114
Fishman, Joshua A. 49
Flint, Peter W. 50
Fox, Harry 93
Fraade, Steven D. 52, 91, 93, 96
Frahm, Eckart 18, 22, 35
Frame, Grant 20, 36, 38
Frede, Dorothea 131
Frede, Michael 131
Frey, Jörg 59, 169
Friedman, Shamma 93
Fromm, Hans 136
Fyzee, Asaf A. A. 195

Gadd, Cyril. J. 36
Gaiser, Konrad 137
Gambetti, Sandra 116
García Martínez, Florentino 59, 76, 77
Geller, Markham J. 21, 24, 32
George, Andrew R. 16, 24, 35, 36, 37, 38
Georgi, Dieter 129
Gesche, Petra D. 9
Gill, Christopher 129
Gokçe, Nuri 28
Goodman, Martin 73

Granfield, Patrick 126
Grant, Robert M. 126
Greenfield, Jonas C. 63
Greidanus, Sidney 85
Griffith, Sidney H. 174, 179, 182
Gurney, Oliver R. 14–16, 24
Guthrie, William K.C. 137

Hackl, Johannes 34, 37
Haggag, Mona 106
Halm, Heinz 190, 192
Halpern-Amaru, Betsy 74, 78, 79, 80–81
Hanegraaff, Wouter J. 121, 122
Hansen, Mogens Herman 112
Harms, Wolfgang 136
Hartog, Pieter B. 48, 50
Haubold, Johannes 33, 35
Hauptman, Judith 91, 93
Heckel, Waldemar 107
Heeßel, Nils P. 14
Heffron, Yagmur 12
Hegermann, Harald 114, 116, 117
Heimerdinger, Jane W. 25
Helmig, Christoph 129
Hempel, Charlotte 57, 66
Hengel, Martin 53, 109–110, 114
Henrichs, Albert 126
Henten, Jan Willem van 112
Hernández-Campoy, Juan Manuel 47
Hezser, Catherine 52, 55
Hilhorst, Anthony 79–80
Himmelfarb, Martha 72
Hirsch-Luipold, Rainer 128, 131, 169
Hirshman, Marc (Menahem) 86–87, 91, 94, 96, 97, 98
Hollenberg, David 194
Honigman, Sylvie 112
Horovitz-Rabin 86, 87
Horst, Pieter W. van der 112
Hourani, George F. 187
Hoyland, Robert G. 52, 58, 182
Hudson, Michael 37
Huehnergard, John 22, 25
Hulin, Peter 15–16, 24
Hunger, Hermann 9, 17, 19–20, 23–24, 25, 33

Inwood, Brad 136

Jacobus, Helen R. 63
James, Patrick 52
Janse, Mark 52, 57
Jastrow, Marcus 94
Jean, Cynthia 32
Joannès, Francis 29
Jones, Alexander 33
Jong, Albert de 122, 123, 134, 135
Joosten, Jan 50, 59
Jungmann, Josef A. 126
Jursa, Michael 28–29, 30, 34, 37

Kahana, Menahem 96
Kaiser, Otto 129
Kampen, John 77
Karmann, Thomas R. 129
Kasher, Aryeh 110, 112, 117
Kelly-Holmes, Helen 48, 52–53
Kersel, Morag 13
Kessler, Nadine 59
King, Karen L. 122, 127
Kingsley, Peter 135–136
Kippenberg, Hans G. 122, 123, 130, 135
Kister, Menahem 98
Kleber, Kristin 34
Klemm, Verena 190
Knibb, Michael 72, 75–76
Knoppers, Gary N. 58
Kokin, Daniel Stein 91
König, Jason 13
Kooten, George H. van 73, 80, 145, 146
Kraft, Robert A. 59
Krämer, Hans J. 137
Kugel, James 79
Kugler, Robert A. 50
Kutscher, Edward Y. 49

Laato, Antti 79
Laks, André 131
Lambdin, Thomas 132
Lambert, Wilfred G. 21, 25, 35, 36, 37
Lamberton, Robert 130
Lanfranchi, Giovanni 33
Lange, Armin 77
Larson, Erik 59
Lauterbach, Jacob Z. 86, 88
Layton, Bentley 132

Leão, Delfim F. 106, 107, 110, 112
Leicht, Reimund 63
Lemaire, André 99
Lenzi, Alan 10, 11, 19, 32
Lévy, Carlos 85
Lewy, Hans (Yohanan) 98
Lieberman, Saul 92, 93, 94, 96, 101
Lieberman, Stephen 18, 19–20, 22–23, 25, 26
Lim, Timothy H. 50
Limor, Ora 176
Lion, Brigitte 32
Lipschits, Oded 58
Littré, Emile 136
Livingstone, Alasdair 16, 19, 22
Lloyd, Seton 28
Long, Anthony A. 152–153
Lucas, Ceil 48
Lüderitz, Gert 112
Luhrmann, Tanya M. 121, 122, 133

Maas, Michael 182
Machinist, Piotr 27
MacMullen, Ramsay 54
Madelung, Wilferd 189–190, 193, 196
Mansfeld, Jaap 136
Marasco, G. 126
Marcus, Ralph 110, 115, 117
Marlow, Louise 171
Marsola, Mauricio Pagotto 127
Martens, John W. 85
Martin, Luther H. 123
Mason, Steve 109, 110, 114
Maul, Stefan M. 14–16, 17
Mayer, Werner R. 32
Mazur, Zeke 127
McAuliffe, Jane D. 171
McEwan, Gilbert J. P. 34
McPherrran, Mark L. 155, 157
Meacham, Tirzah 93
Meissner, Burkhard 134
Mélèze Modrzejewski, Joseph 110, 111, 114, 115, 117
Metso, Sarianna 47, 62
Meyer, Ben F. 136
Michel, Cécile 32
Millar, Fergus 58

Morenz, Ludwig 52
Mullen, Alex 52, 58
Muraoka, Takamitsu 47

Najman, Hindy 47, 62, 73, 80, 85
Narbonne, Jean-Marc 127
Navascués, Patricio de 128
Naveh, Joseph 57
Nebe, G. Wilhelm 53
Neher, Martin 129
Neugebauer, Otto 10, 31, 35, 37
Nickelsburg, George W. E. 72
Nicklas, Tobias 129
Nutton, Vivian 136

Oelsner, Joachim 35
Oeming, Manfred 58
Oikonomopoulos, Katerina 13
Olman, Arye 172–173
Opsomer, Jan 128, 131
Ossendrijver, Mathieu 35
Oudshoorn, Jacobine G. 55

Parker, Victor 107
Parpola, Simo 18–19, 20, 26–27
Paul, Shalom M. 59
Pepin, Jean 123
Perdue, Leo G. 103
Pereira, Maria Helena da Rocha 108
Petersen, Anders K. 145
Pietikäinen, Sari 48, 52–53
Pinches, Theophilus G. 34
Pirngruber, Reinhard 33, 34
Places, Édouard Des 128, 136
Poonawala, Ismail 195
Popović, Mladen 48, 54, 55–57, 59, 60, 61–65
Porter, Barbara Nevling 28
Postgate, J. Nicholas 21, 26, 27
Potter, Paul 129
Press, Gerald A. 153
Price, Jonathan J. 52, 58

Rabin, Chaim 49
Radner, Karen 9, 22, 23, 24
Ratzon, Eshbal 61
Reade, Julian E. 22, 24
Reale, Giovanni 137

Reed, Annette Yoshiko 103
Reed, Stephen 60
Reese, James M. 129
Reeve, Christopher D. C. 156
Reinhold, Meyer 108, 109, 115
Reinprecht, Christoph 112
Reisner, George 35, 36
Rendsburg, Gary A. 47
Reynolds, Gabriel S. 172, 174, 182
Richey, Matthew 59
Riedweg, Christoph 136
Roberts, Alexander 124, 125
Robson, Eleanor 8, 9, 12, 13, 14, 17, 22, 23, 26, 27, 28–31, 35, 64, 65
Rodrigues, Nuno S. 106, 108, 117
Roig Lanzillotta, Lautaro 122, 123, 128–129, 134
Rollinger, Robert 29, 33
Rose, Jonathan 26
Rössler, Paul 47
Rouillard, Pierre 29
Rozen, Minna 110
Ruberg, Uwe 136
Ruggles, Clive N. 33
Ruiten, Jacques T. A. G. M. van 72, 73–74, 76, 77, 78, 79, 81
Ruppel, Walter 112
Rutkowska, Hanna 47
Rutz, Matthew 13
Ryholt, Kim 13

Sachs, Abraham J. 33
Sáez, Andrés 128
Sanders, Ed P. 129, 136
Sanders, Seth L. 48, 63, 64
Sänger, Patrick 112–113, 114
Schendl, Herbert 47
Schiffman, Lawrence H. 47, 59
Schneider, Helmuth 137
Schniedewind, William H. 47, 49, 50–51
Schoonover, Myles 48, 63–64
Schorch, Stefan 52
Schuller, Eileen 47, 62
Schweizer, Eduard 130
Schwind, Fritz F. Von 87
Sedley, David 131
Segal, Michael 77, 80
Selz, Gebhardt 13

Sharples, Robert W. 131
Sievers, Wiebke 112
Silva, Maria de Fátima 108
Simmel, Georg 121, 122, 133
Sjöberg, Åke W. 25
Smelik, Willem F. 49, 52, 57–58, 91, 93
Smith, Jonathan Z. 127
Smith, Nicholas D. 156
Sokoloff, Michael 63
Sorabji, Richard 131
Sousa, Rogério 106
Spar, Ira 35, 36, 37
Spek, Robartus J. van der 34
Spolsky, Bernard 49
Stackert, Jeffrey 28
Staden, Heinrich von 129
Steele, John 33, 35
Steinkeller, Piotr 22, 25, 37
Stemberger, Günter 93
Sterling, Gregory E. 85
Stevens, Kathryn 10, 11, 12–13, 14, 17, 30, 31, 35, 65
Stewart, Jon B. 131
Stock, Brian 48
Stone, Adam 12
Strack, Herman L. 93
Stroumsa, Guy G. 122, 123, 125, 130, 135, 176
Swain, Simon 52, 57
Szlezák, Thomas A. 137

Tenu, Aline 29
Thompson, Dorothy J. 52
Thür, Gerhard 112
Tigchelaar, Eibert J. C. 47, 50–51, 58, 59, 60, 73
Tigerstedt, Eugène N. 137
Topper, Kathryn 145, 158
Totelin, Lawrence M. V. 134
Tov, Emanuel 47, 50, 51, 59–61, 62
Tröster, Manuel 107
Troyer, Kristin De 60
Truschnegg, Brigitte 29
Tso, Marcus K. M. 53
Turner, John D. 127
Tzoref, Shani 47

Ulrich, Eugene 50
Unger, Dominic J. 124
Urban, Hugh B. 121, 122, 133

Vandenberghe, Marijn 48, 63–64
Vanderhooft, David 58
VanderKam, James C. 72, 76, 77, 80
Vaux, Roland de 54
Veldhuis, Niek 9
Vermaseren, Maarten J. 126, 128, 136
Villard, Pierre 22

Waerzeggers, Caroline 12, 29, 64
Wagensonner, Klaus 13
Waldstein, Michael 133
Walker, Paul E. 188–190, 193, 196
Wassén, Cecilia 60
Wasserstein, David J. 52, 58
Watt, Jan G. van der 169
Wear, Andrew 136
Weidner, Ernst 36
Weiher, Ernst von 32
Weinfeld, Moshe 98
Weitzman, Steven 47, 49, 50, 53, 78
Werman, Cana 72
West, Martin L. 131
Whitehead, David 112
Wiesehöfer, Josef 34
Wilk, Florian 75
Williams, Michael A. 121, 125, 127
Williamson, Hugh G. M. 72
Wise, Michael O. 52, 54, 55–57, 59, 65
Wiseman, Donald J. 18, 21, 25
Wisse, Frederik 133
Wolff, K. 121
Wolfson, Elliot R. 121
Woodruff, Paul B. 156
Woolf, Greg 13
Worthington, Martin 12
Wright, David P. 28
Wright, John 129

Yadin, Yigael 60
Yardley, John C. 107

Zellentin, Holger Michael 174, 176
Zimmermann, Ruben 169
Zólyomi, Gábor 23
Zuckerman, Constantine 112, 114

Sources Index

Cuneiform sources
Astronomical Cuneiform Texts (Neugebauer)
– no. 135U rev. 12–16 23
– no. 135U rev. 12–16 10
Babylonische und Assyrische Kolophone (Hunger)
– no. 193 rev. 22–27 15
– no. 297A 20
– no. 299 20
– no. 325 23
– no. 498 12–17 20
Letters from Assyrian and Babylonian Scholars (Parpola)
– no. 202 obv. 5–13 19
Literary Texts (Wiseman and Black)
– no. 170 21
– no. 188 rev. ii 5'–7 21
The Sultantepe Tablets, Volume I (Gurney and Finkelstein)
– no. 38 rev. ii 11–13 10, 15
– no. 38 rev. ii 16–18 10, 15
– no. 301 rev. ii 11'–iii 12' 15

Bible
Old Testament
– Genesis
  – 1–11 74n4
– Exodus
  – 19:2 86
– Leviticus
  – 14:34 98
– Deuteronomy
  – 20 98
  – 20:10 98
  – 20:10–11 96, 98
  – 20:10–14 98n32
  – 20:10–18 95
  – 20:18 92, 95, 101
  – 20:19 96
  – 21:10 102
  – 27 101
  – 27:2 100
  – 27:2–3 93
  – 27:2–8 89, 90
  – 27:3 101
  – 27:4 97, 101
  – 27:8 90, 91, 93, 94n23, 96–100
  – 27:84 97
  – 27:12–28:68 91
  – 27–28 90, 91
– Joshua
  – 4:20 92
  – 4:3 90, 92
  – 4:5 92
  – 4:8 90
  – 4:24 89
  – 8:32 96, 97
  – 8:30–35 89, 90
– 2 Maccabees 129n42
– Psalms
  – 55:24 94n23
  – 138:4 88n10
  – 147:19–20 87n10
– Wisdom of Solomon
  – 2:2–4 129n42
  – 8:19–20 129n42
  – 9:15 129n42
– Isaiah
  – 33:12 100
  – 45:19 87
  – 60:12 100
– Habakkuk
  – 1:17 50
  – 2:16 50
  – 3:3–6 88n10
New Testament
– Matthew
  – 3:17 164n15
  – 9:34 164n15
  – 10:26 123n15
  – 12:24 164n15
  – 17:5 164n15
  – 27:46 164n15
  – 27:50 164n15
– Mark
  – 1:11 164n15
  – 3:22 164n15
  – 9:7 164n15

- 15:34 164n15
- 15:37 164n15
– Luke
  - 3:22 164n15
  - 9:35–36 164n15
  - 11:15 164n15
  - 23:46 164n15
– John
  - 1:9 146
  - 1:12 159
  - 1:18 146
  - 1:39 161
  - 2:3–5 150
  - 2:4 157, 161, 162
  - 2:7–8 162
  - 2:8 146
  - 2:8–9 157
  - 2:11 157
  - 2:18 159, 168
  - 2:18–21 157
  - 2:23 157
  - 3:2 158, 159
  - 3:29 164n15
  - 4 145
  - 4:6 161
  - 4:21 159, 161
  - 4:23 159
  - 4:52–53 158, 161
  - 4:54 158
  - 5:16–18 160
  - 5:18 159
  - 5:25 161
  - 5:31 160
  - 5:36–37 160
  - 5:37 160, 164n15
  - 6 145
  - 6:2 158
  - 6:11 158
  - 6:14 158
  - 6:26–27 158
  - 6:30 158
  - 6:32 158
  - 6:55 158
  - 6:60 167
  - 7:1 159
  - 7:2–5 162
  - 7:2–10 158
  - 7:2–14 150
  - 7:6 161, 162
  - 7:8 161, 162
  - 7:10–14 162
  - 7:13 158
  - 7:14–18 164n15
  - 7:15 154n9
  - 7:19 159
  - 7:20 164n15
  - 7:25 159
  - 7:30 159, 161, 162
  - 7:33 162
  - 7:53–8:11 154n9
  - 8:1–5 162
  - 8:6 154
  - 8:6–8 162
  - 8:8 154
  - 8:12–18 162
  - 8:20 161, 162
  - 8:37 159
  - 8:38–42 164n15
  - 8:40 159
  - 8:41–42 159
  - 8:48–49 164n15
  - 8:49 164n15
  - 8:50 164n15
  - 8:59 167
  - 9:16 158
  - 10:3–5 164n15
  - 10:6 168
  - 10:14–18 164n15
  - 10:16 164n15
  - 10:20–21 164n15
  - 10:27 164n15
  - 11:1–7 162
  - 11:6 158
  - 11:9 161
  - 11:47–48 159
  - 11:49–53 159
  - 11:53 159
  - 12 146
  - 12:18 158
  - 12:20–28 163
  - 12:23 161
  - 12:27 161
  - 12:27–33 164n15
  - 12:28 164n15

New Testament (continued)
- 12:30 164n15
- 12:33 157, 159, 168
- 12:35 163
- 12:36 167
- 12:37 159
- 12:42–43 159
- 12:44–50 163
- 13:1 161, 163
- 13:1–10 163
- 13:5–11 167
- 13:10–11 169
- 13–17 146, 166
- 14:27 168
- 14:31 163
- 15:3 167, 169
- 15:27 160
- 16:2 161
- 16:4 161
- 16:21 161
- 16:25 162, 163, 165
- 16:25–28 168
- 16:25–30 169
- 16:29–30 168
- 16:32 161
- 16:33 168
- 17:1 161, 163
- 18:19–20 167
- 18:28 154
- 18:32 157, 159, 169
- 18:37 164n15
- 19:4 161
- 19:27 161
- 19:30 146
- 19:31–34 154
- 20:17 159
- 20:30 159
- 20:30–31 169
- 20:31 159
- 21 145
- 21:24 160
- 1 Thessalonians 5:23 131n45

Old Testament Pseudepigrapha
1 Enoch 72n1
- 81:6 76
Jubilees
- 1:27–28 75

- 20 75
- 2–10 74n4
- 2:26 75
- 2:29 75
- 4 75
- 4:18 76
- 4:22 76
- 4:15 81
- 4:17 75, 76
- 4:17–19 77n11
- 4:17–26 75–76
- 4:18–19 76
- 4:21–22 76
- 4:21–23 77n11
- 4:23 76
- 4:23–24 76
- 5:1 81
- 6:13 75
- 7:21 81
- 7:23–24 81
- 7:38 76
- 7:38–39 77n11, 81
- 7:39 76
- 8:11 76
- 10:13 77
- 10:13 77n11
- 10:14 77
- 11:4–5 82
- 11:8 82
- 11:16–17 77
- 12:16–18 82
- 12:25–27 77
- 12:25–27 49
- 12:27 77n11, 78
- 15:28 75
- 15:31–32 77n11
- 19:14 78
- 19:24 81
- 21 78
- 21:10 77n11, 78
- 22:16–22 72
- 30:11 75
- 32 75
- 39:5–7 79
- 39:6 77n11
- 45:14–16 79
- 45:16 77n11
- 47:9 79

Sources Index

Oxyrhynchus Papyri
POxy 1007 60
POxy 3522 60

Dead Sea Scrolls an Related Texts
1Q70/1Q70bis 59n41
4Q186 65
4Q201 (4QEna ar) 59n41
4Q216 72n1
4Q249 61
4Q266 10 i 3, 53
4Q298 61
4Q318 63
4Q338 59n41
4Q350 59n41
4Q460 9, 59n41
4Q464 49
4Q477 2 ii 5, 59
4Q518/4Q519 59n41
7Q-Arch, 2 heb/ ar 49
Genesis Apocryphon
– 19:24–25 76
Isaiah Scroll 49, 53
Pesher Habakkuk 50, 53
Cairo Genizah
– Damascus Document
  – 14:8–10 53
  – 16:2–4 72n1
Naḥal Ḥever
– 1/8ḤevXIIgr 60
– Bar Kokhba letters,
   P.Yadin 52 49

Nag Hammadi Codices
Gospel of Truth
– [NHC I,3]
  – 18.11–15 133n54
Apocryphon of John
– [NHC II,1]
  – 1–4 133n50
  – 31.29–31 133n51
– [BG]
  – 75.15–20 133n52
Gospel of Thomas
– [NHC II,2]
  – 2 133
  – 50 132

Ancient Jewish Writers
Josephus
– *Against Apion*
  – 1.176–182 108
  – 1.186–204 109
  – 1.189 116
  – 1.190–192 109
  – 11.52–113 126n26
– *Jewish Antiquities*
  – 2.236 79n17
  – 11.304–346 110
  – 11.337–339 111
  – 12.107–108 115–116
  – 14.117 117
Philo
– *Life of Moses*
  – 1:8–24 79n17

Rabbinic Literature
Babylonian Talmud
– Soṭah
  – 35b–36a 98n31
  – 35b–36a 101
Deuteronomy Rabbah
– 5.14 98
Jerusalem Talmud
– Shevi'it
  – 6.1 98
– Soṭah
  – 7.5 100
Leviticus Rabbah
– 17.5–6 98
Mekhilta Deuteronomy
– Deuteronomy
  – 27:8 96
midrasim
– Mekhilta deRabbi
   Ishmael
  – Baḥodesh 1 86–87
Mishnah Megillah
– 4:10 173
Mishnah Soṭah
– 7:1–5 91
– 7:5 90–92, 98n31
Siffre Deuteronomy
– Deuteronomy
  – 33:2 87

Tosefta Soṭah
– 8:6–7   92–95

Graeco-Roman Literature
Cleanthes
– SVF I 538   136n70
Dio Cassius
– 73.16   126
– 79.1   126
Diodorus of Sicily
– *Bibl. Hist*
   – 40.3   108–109
Exagoge
– 36–38   79n17
Heraclitus
– B 2 DK   135n65
– B 4 DK   135n66
– B 5 DK   135n66
– B 11 DK   135n65
– B 17 DK   135n65, 135n66
– B 39 DK   165, 166
– B 41 DK   135n66
– B 54 DK   135n65
– B 93 DK   147
– B 101 DK   135n67
– B 123 DK   135n65
Herodotus
– *Histories*
   – 2.104.2–3   108
Hesiod, *Works and Days* 25–26   135n62
Hippocratic nomos   136
– *Law* 5   136n68
Hippocratic Oath   134
– 5–6   134n61
Historia Augusta
– 8.1–2   126
Irenaeus
– *Against all Heresies*
   – 1.31.4   124
Juvenal
– 6.550–52   126
Livy
– 39.8–18   125
Lucretius: 128
– *De rerum natura* 1.69–72   128n33
Philo
– *Legum allegoriae*
   – 1.105–108   166

– *Quis rerum divinarum heres sit*
   – 213–214   166
Plato
– *Apology*
   – 20D–23C   153
   – 20E   160
   – 20E–21A   153, 154
   – 21A   160
   – 22E–23A   159
   – 23A–B   153
   – 23B–C   153
   – 23E–24A   159
   – 24B–C   148
   – 26B   148
   – 30A–31A   159
   – 31B–C   153
   – 31C–D   164n15
   – 31D   148, 153, 164n15
   – 33B   153
   – 40A   155, 164n15
   – 40A–C   148, 155n10, 161, 163n15
   – 40B   162
   – 41C–D   149, 155n10
   – 41D   163n15
   – 42A   149, 161, 163
– *Cratylus*
   – 398B–C   152
   – 405A   155
   – 405C   155
– *Epinomis*
   – 988A   156n12
– *Epistle*
   – II 314   136n69
   – VII 344B   136n69
– *Euthydemus*
   – 272E–373A   150
– *Euthyphro*
   – 3B   147, 164n15
– *Laws*
   – 654A   155
   – 738B–C   155n12
   – 759C–D   155n12
   – 828A   155n12
   – 923A   155
   – 968E 2–5   137n72
– *Phaedo*
   – 58B–C   154
   – 59D–60A   161

- 60D 154
- 61B 154
- 64D 158
- 65D–66B 146
- 65D–E 158
- 66B–C 158
- 69C–D 136n69
- 81B 146, 158
- 84A–B 146, 158
- 84E–85B 155
- 89C 161, 163
- 109E 146
- 115A 163
- 116A 163
- *Phaedrus*
  - 24B–D 150
  - 243B 137n72
  - 250A–251A 136n69
  - 265B 155
- *Protagoras*
  - 343A–B 155
- *Republic*
  - 2.13, 372B–373A 145
  - 427B–C 155
  - 427C 155–156n12
  - 496A–C 151
  - 540A–C 156n12
  - 540B–C 156n12
- *Symposium*
  - 174A–B 149
  - 174D–175C 162
  - 174D–E 149
  - 175A 149
  - 175B 150
  - 175C 150
  - 197A 155
  - 209E–210A 136n69
  - 217D–219D 158
- *Theaetetus*
  - 151A 151
  - 152C 137n72
- *Timaeus*
  - 42D 128n34
Plutarch
- *De genio Socratis*
  - 581A–B 152
- *De Pythiae oraculis*

- 404D–E 166
- 406C 166
- 406D–F 166
- 407A–B 166
- 407C–D 166, 168
- 408B 166, 168
- *De stoicorum repugnantiis*
  - 1035A–B [=SVF II, 42] 136n70
- *Fragm.*
  - 202 167
- *Platonic Questions*
  - 999D–E 152
Porphyry
- *De abstinentia*
  - 2.26 108
Proclus
- *In Tim.*
  - 1.303.27–304.3 128n36
Pseudo-Aelius Aristides
- *Ars rhetorica*
  - 2.2.1.7 167
  - 2.3.1.6 167
  - 2.5.1.2 167
Simonides
- fr. 542.ll-6 Page 135
Solon
- fr. 20 West 135
Xenophon
- *Memorabilia*
  - 1.1.1–4 148

Patristics
Clement of Alexandria
- *Exc. Ex Theod.* 78 133n57
Irenaeus 123, 124, 131
- *Adv. Haer.* 5.6.1 131n45
- *Adv haer*. Prol. 2 123n16
Pseudo-Hippolytus 124, 125
- *Refutatio praef* 5 124
- *Refutatio praef.* 7–8 125
Tertullian 125, 131
- *Apol.* 9.9 126n25
- *De carne Christi* V, 4 131n47

Qur'ān
2:75 176
2:79 172

3:7 186
3:58 191
3:67 176n8
3:75 180
3:75–80 173n9
3:78 172, 174
4:46 174
5:44 176
5:63 176, 177
5:82 177, 181
5:82–9:31/34 181n35
6:7 171
6:91 171, 173
7:154 171
9:30–34 175, 177, 181–182
9:31 176
9:32 176, 180
9:34 176
10:94 171
24:15 174
26:195–197 171
35:28 171n3
52:3 171
59:22 171

www.ingramcontent.com/pod-product-compliance
Lightning Source LLC
Chambersburg PA
CBHW031816220426
43662CB00007B/674